Power Play

Power Play

Edward J. Barbeau

University of Toronto

Published by
THE MATHEMATICAL ASSOCIATION OF AMERICA

SPECTRUM SERIES

The Spectrum Series of the Mathematical Association of America was so named to reflect its purpose: to publish a broad range of books including biographies, accessible expositions of old or new mathematical ideas, reprints and revisions of excellent out-of-print books, popular works, and other monographs of high interest that will appeal to a broad range of readers, including students and teachers of mathematics, mathematical amateurs, and researchers.

Mathematical Association of America
1529 Eighteenth Street, NW
Washington, DC 20036
800-331-1MAA   FAX 202-265-2384

To
Edward Orlando Oscar Arthur (1878–1948)
Alexander Joseph Gargaro (1993–)
Maxwell Edward Gargaro (1996–)

Preface

When I was young, my grandfather introduced me to the number 142857, which has the property that its first six multiples involve the same six digits in the same cyclic order. This was the beginning of a fascination with numbers, which I continue to nurture on public transportation and in queues. The natural numbers exhibit a myriad of patterns and properties, some of which can be readily established by an alert school student and others of which reflect deep properties that can be illuminated, if at all tractable, only by the most advanced techniques.

Numbers, like the stars in the heavens or the natural world around us, are everyone's birthright. Along with geometrical figures and puzzles (logical or topological), they are accessible to amateurs. Just as the stargazer, naturalist or rockhound can ply an avocation without advanced study in astronomy, biology, geology or paleontology, so one can happily play with numbers in ignorance of any but the most basic mathematics. However, the comeliness of the number system have probably induced more than one child to explore more deeply and adopt mathematics as a vocation.

As L.E. Dickson's encyclopædic *History of the theory of numbers* demonstrates, the appreciation of singular properties of numbers reaches back to antiquity. The Babylonians knew about pythagorean triples and the Greeks about primes and perfect numbers. The Fibonacci numbers have been studied for almost 800 years and the Pascal triangle was known long before Pascal saw the light of day. The introduction of Arabic numeration brought in its train all sorts of digital oddities.

This book focuses on powers of numbers. This is not unduly narrow. The most notorious problem of all time, the Fermat conjecture that no positive nth power for $n \geq 3$ can be written as the sum of two positive nth powers, falls within this category. A few of the results recorded here are quite recent, so that I hope that this material will be of interest to the experienced mathematician as well as the novice.

Strict amateurs who ignore the exercises and notes will find that a background of elementary mathematics with basic high school algebra will carry them through. However, there are other audiences to whom this monograph is addressed.

(a) **Secondary students** should read this book with a calculator handy. Besides checking the results given, they may want to explore and discover patterns of their own. Advanced secondary students should try to prove some of the results given. The exercises, keyed to assertions in the text, will indicate what they can credibly tackle. The notes will provide further details.

(b) **College students** might follow up some of the references to the literature and think about the structural and theoretical issues behind the patterns. Primitive pythagorean triples with a suitable operation form a group; is this group cyclic? The solutions of Pell's equation constitute groups of units in field extensions of finite degree over $\mathbf{Q}$; the Dirchlet unit theory indicates the structure.

(c) **Secondary teachers** will be able to construct units that will supplement the regular curriculum. Many of the arithmetic equations can be verified through factorization techniques. Searching out and verifying special relationships among larger numbers will require skills in numerical analysis and the use of a calculator or computer to a high degree of efficiency.

(d) **College teachers** will find in this book material for student projects and essays.

Through the body of each chapter, certain topics and assertions are keyed with a boldface integer in parentheses. These refer to entries in both **Exercises on the Notes** and **Notes** sections at the conclusion of the chapter. I have put the Exercises on the Notes first, in order to encourage the reader to try to establish the keyed result first before looking at the discusion, details of proof and references to the literature in the Notes themselves. After the notes, additional exercises and solutions are provided.

I am indebted to the following students for searching various journals for material: Belal Ahmad, Joe Callaghan, Christine Chu and Dina Sum. In addition, Cyrus Hsia and Joseph Deu Ngoc checked the numerical calculations for the cubic Pell's equation. The support of my wife, Eileen, during this project has been invaluable.

Contents

CHAPTER I

Odd Integers and Squares

The squares of the positive whole numbers are $1, 4, 9, 16, 25, 36, \dots$. Look at the differences between successive squares. They are the odd numbers: $3, 5, 7, \dots$. If we add successive odd numbers, we get all the squares in order:

$$1 = 1; \quad 1 + 3 = 4; \quad 1 + 3 + 5 = 4 + 5 = 9;$$

$$1 + 3 + 5 + 7 = 9 + 7 = 16; \quad 1 + 3 + 5 + 7 + 9 = 16 + 9 = 25.$$

From what we have observed so far, we might surmise that the sum of the first 6 odd numbers is the square of 6, the sum of the first 7 odd numbers is the square of 7, and so on. Would you wager that the sum of the first million odd numbers is the square of 1,000,000? If so, you would win.

There is a nice way of seeing that, for any number n, the sum of the first n odd numbers is the square of n. Arrange n^2 dots in a square array, n dots to a side. By partitioning the array into L-shaped sections (called *gnomons*), we can see at once that the number of dots is the sum of the first n odd positive integers. The diagram illustrates this for $n = 10$:

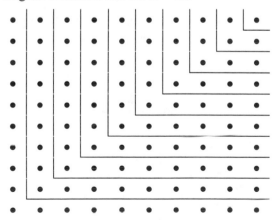

When one has a nice mathematical result, it is natural to see whether it is part of a larger pattern. One is rarely disappointed.

We can think of the odd numbers as what is left over when we cross out the even numbers, i.e., every second number. The squares then are the running totals of the remaining numbers. Instead of crossing out overy other number, cross out every third number and then take a running total:

1	2	3	4	5	6	7	8	9	10	11	12	13	14	15	16	17
1	3		7	12		19	27		37	48		61	75		91	108

In the new sequence, cross out every second number and then take a running total:

1	3	7	12	19	27	37	48	61	75	91	108
1		8		27		64		125		216	

We get the sequence of cubes! The reader may wish to check that the cubes continue to appear if we extend the sequences.

Can this be generalized further? We have a process yielding squares which when generalized gives cubes. There must be an analogous way of generating fourth powers. Do you see how?

Again, start with the sequence of natural numbers, but this time striking out every fourth one. We then strike out every third number of the sequence of running totals, and then every second one of the next sequence of running totals. The end result is

1	2	3	4	5	6	7	8	9	10	11	12	13	14	15	16
1	3	6		11	17	24		33	43	54		67	81	96	
1	4			15	32			65	108			175	216		
1				16				81				256			

The last sequence consists of the fourth powers of all the positive integers.

The reader can discover how the higher powers of the integers can be generated **(1)**.

Another way of generalizing the fact that the sum of the first n odd integers is the square of n is illustrated by the following set of tables. Each is assumed to continue down and across indefinitely.

TABLE I

1	1	1	1	1	1	1
0	1	2	3	4	5	6
0	1	3	6	10	15	21
0	1	4	10	20	35	56
0	1	5	15	35	70	126
0	1	6	21	56	126	252

TABLE II

1	1	1	1	1	1	1
2	3	4	5	6	7	8
2	5	9	14	20	27	35
2	7	16	30	50	77	112
2	9	25	55	105	182	294
2	11	36	91	196	378	672

TABLE III

1	1	1	1	1	1	1
0	3	8	15	24	35	48
0	5	27	84	200	405	735
0	7	64	300	1000	2695	6272
0	9	125	825	3675	12740	37044
0	11	216	1911	10976	47628	169344

TABLE IV

1	1	1	1	1	1	1
1	4	9	16	25	36	49
1	9	36	100	225	441	784
1	16	100	400	1225	3136	7056
1	25	225	1225	4900	15876	44100
1	36	441	3136	15876	63504	213444

The first columns of Tables I and II are chosen as indicated. Every other column in the tables arise from these two according to the rules:

1. Each column of Table I after the first is a running total of the column before. (Table I is Pascal's triangle on a slant.)
2. Each column of Table II after the first is a running total of the column before.
3. Each column of Table III consists of the products of the corresponding elements in the same columns of Tables I and II.
4. Each column of Table IV is a running total of the corresponding column of Table III.

Now let us make some observations. The phenomenon that the successive sums of odd numbers are squares is reproduced in two places, in columns 2 and 3 of Table II and in the second columns of Tables III and IV. Moreover, it turns out that all the numbers in Table IV are squares, and, indeed, each column of Table IV is the sequence of squares of the terms in the *next* column of Table I (2).

There is another interesting fact to emerge from these tables. Look at column 3. In Table I, this column consists of the successive integers; in Table II, it consists of their squares. Thus, in Table III, the third column consists of the cubes of the positive integers, and in Table IV, the running totals of these cubes. These totals are all squares. Thus, we find

$$1^3 = 1^2$$
$$1^3 + 2^3 = (1 + 2)^2$$
$$1^3 + 2^3 + 3^3 = (1 + 2 + 3)^2$$
$$1^3 + 2^3 + 3^3 + 4^3 = (1 + 2 + 3 + 4)^2$$
$$1^3 + 2^3 + 3^3 + 4^3 + 5^3 = (1 + 2 + 3 + 4 + 5)^2$$

and in general, for any positive integer n,

$$1^3 + 2^3 + 3^3 + \cdots + n^3 = (1 + 2 + 3 + \cdots + n)^2.$$

This equation can be established in a variety of different ways **(3)**.

A link between the sum of the first n cubes and the sum of consecutive odd integers can be illustrated in the following triangular array of numbers:

$$1$$
$$3 \qquad 5$$
$$7 \qquad 9 \qquad 11$$
$$13 \qquad 15 \qquad 17 \qquad 19$$

The sum of the terms in the nth row is n^3 (do you see why?), and the sum of all the terms in the first n rows is $(1 + 2 + \cdots + n)^2$ **(4)**.

If we put the first n^2 odd numbers in a square, as illustrated for $n = 4$, then the diagonal sum, the first row sum and the sum of all the entries are of interest.

$$
\begin{array}{cccc}
1 & 3 & 5 & 7 \\
9 & 11 & 13 & 15 \\
17 & 19 & 21 & 23 \\
25 & 27 & 29 & 31
\end{array}
$$

In general, the first row sum is n^2, the diagonal sum is n^3 and the sum of the whole array is n^4 **(5)**.

But let us return to the sums of cubes. Another way of stating the fact about the sum of the first n cubes is that $\{1, 2, \ldots, n\}$ is a set of numbers with the property that *the sum of the cubes of the numbers in the set is equal to the square of their sum.* Are there any other such sets of numbers?

The perhaps surprising answer is *yes*, if we allow repetitions of numbers in the set. Before reading on, try to find some other sets of integers with the stated property.

Here are some: $\{2,2\}$, $\{3,3,3\}$, $\{4,4,4,4\}$, and so on. There is a very general way of generating other sets of numbers with the property. Choose any positive integer. We will take the number 24 as an example. In a column, write down each of its positive divisors; beside each divisor, write down a list of *its* divisors. Then note how many divisors each divisor has in the third column. Finally, in the fourth column, write the cubes of the elements in the third column. Now total up the elements in the third and fourth columns. For 24, the arrangement looks like this:

divisor	its divisors	number	its cube
1	1	1	1
2	1, 2	2	8
3	1, 3	2	8
4	1, 2, 4	3	27
6	1, 2, 3, 6	4	64
8	1, 2, 4, 8	4	64
12	1, 2, 3, 4, 6, 12	6	216
24	1, 2, 3, 4, 6, 8, 12, 24	8	512
Totals		30	900

Thus $1^3 + 2^3 + 2^3 + 3^3 + 4^3 + 4^3 + 6^3 + 8^3 = (1 + 2 + 2 + 3 + 4 + 4 + 6 + 8)^2$. Try this with other numbers. It works every time! **(6)**

I am indebted to E. Wang of Wilfrid Laurier University in Waterloo, Ontario for bringing to my attention the work of the Chinese mathematician Yen-Kun Yin. Yin has two ways of constructing new sets of integers whose cube sums are equal to the squares of their sums **(7)**.

The best way to describe his first technique is to give some specific examples:

$\{1,2\}$ $\{1,2,3,4\}$ $\{1,2,2,4\}$ $\{4,2,2,4\}$ $\{1,2,3,4,5,6\}$ $\{1,2,3,4,2,6\}$

$\{1,2,4,4,2,6\}$ $\{6,2,4,4,2,6\}$ $\{1,2,3,4,5,6,7,8\}$ $\{1,2,3,4,5,6,2,8\}$

$\{1,2,3,4,4,6,2,8\}$ $\{1,2,6,4,4,6,2,8\}$ $\{8,2,6,4,4,6,2,8\}$.

In general, if n is a positive integer and $0 \le k \le n$, then the set

$\{1, 2, 3, \ldots, 2(n-k), 2k, 2(n-k+1),$

$$2(k-1), 2(n-k+2), \ldots, 4, 2(n-1), 2, 2n\}$$

has the desired property.

Yin's second method is to begin with the two sets $\{a_1, a_2, \ldots, a_m\}$ and $\{b_1, b_2, \ldots, b_n\}$ and form a third set consisting of all mn products $a_i b_j$ ($1 \le i \le m$, $1 \le j \le n$). For example, the sets $\{1, 2, 3\}$ and $\{1, 2, 2, 4\}$ yield

$$\{1, 2, 2, 2, 3, 4, 4, 4, 6, 6, 8, 12\}.$$

There is a striking equation involving the fourth power of the sum of the first n positive whole numbers and the sums of their fifth and seventh powers:

$$(1^5 + 2^5) + (1^7 + 2^7) = 162 = 2(1 + 2)^4$$
$$(1^5 + 2^5 + 3^5) + (1^7 + 2^7 + 3^7) = 2592 = 2(1 + 2 + 3)^4$$
$$(1^5 + 2^5 + 3^5 + 4^5) + (1^7 + 2^7 + 3^7 + 4^7) = 20000 = 2(1 + 2 + 3 + 4)^4$$

and so on **(8)**.

A relationship similar to that for the sum of the cubes just described involves numbers that are *relatively prime* to a given number. (Two integers are *relatively prime* or *coprime* when their greatest common divisor is 1.) Given an integer n, we begin by writing down all the positive numbers less than n and coprime with n. Then sum these integers and sum their cubes. The table below gives the results for $1 \le n \le 12$.

number	smaller coprime numbers	their sum	their cube sum
1	1	1	1
2	1	1	1
3	1, 2	3	9
4	1, 3	4	28
5	1, 2, 3, 4	10	100
6	1, 5	6	126
7	1, 2, 3, 4, 5, 6	21	441
8	1, 3, 5, 7	16	496
9	1, 2, 4, 5, 7, 8	27	1053
10	1, 3, 7, 9	20	1100
11	1, \ldots, 10	55	3025
12	1, 5, 7, 11	24	1800

For each given value of n,

the sum of $(n/d)^3 \times$ (sum of the cubes of numbers coprime with d) over all divisors d of n is equal to the square of the sum of $(n/d) \times$ (sum of numbers coprime with d) over all divisors d of n.

For example, if $n = 9$, we have that

$$9^3 \times 1 + 3^3 \times 9 + 1^3 \times 1053 = 2025 = 45^2$$
$$= (9 \times 1 + 3 \times 3 + 1 \times 27)^2;$$

if $n = 6$, we have that

$$6^3 \times 1 + 3^3 \times 1 + 2^3 \times 9 + 1^3 \times 126 = 441 = 21^2$$
$$= (6 \times 1 + 3 \times 1 + 2 \times 3 + 1 \times 6)^2;$$

and if $n = 12$, we get

$$12^3 \times 1 + 6^3 \times 1 + 4^3 \times 9 + 3^3 \times 28 + 2^3 \times 126 + 1^3 \times 1800 = 6084 = 78^2$$
$$= (12 \times 1 + 6 \times 1 + 4 \times 3 + 3 \times 4 + 2 \times 6 + 1 \times 24)^2.$$

The reader may have noticed that the sum over the divisors d of n of the terms $(n/d) \times$ (sum of numbers coprime with d) is, in these examples, equal to the sum $1 + 2 + 3 + \cdots + n$. Is this always true? **(9)**

Since we have raised the issue of divisors, we should note a nice characterization of squares. They are the only positive integers which have an odd number of divisors **(10)**. Do you see why this is so? There is an old problem which is based on this fact:

A set of lockers is numbered consecutively $1, 2, 3, 4, \ldots$; all are originally locked. A janitor comes along and unlocks each one in turn. A second janitor then locks each second one, i.e., those numbered $2, 4, 6, \ldots$. A third janitor changes the state of each third locker (those numbered $3, 6, 9, \ldots$) by locking those that are unlocked and unlocking those that are locked. A fourth janitor changes the state of each fourth locker. This continues indefinitely. Which lockers are unlocked at the end? **(11)**

We have been looking at sums of cubes and squares. But there is an interesting result about differences of cubes:

If the difference of two consecutive cubes is a perfect square, then the number squared is the sum of two consecutive squares **(12)**.

Let us look at some numerical evidence:

$$2^3 - 1^3 = 7, \quad 3^3 - 2^3 = 19, \quad 4^3 - 3^3 = 37,$$
$$5^3 - 4^3 = 61, \quad 6^3 - 5^3 = 91, \quad 7^3 - 6^3 = 127$$

are all differences of consecutive cubes; none of these are squares. But $8^3 - 7^3 = 169$, which is the square of 13. And, indeed, $13 = 2^2 + 3^2$ is the sum of two consecutive squares.

We have to go quite a bit further along before we come to the next difference of consecutive cubes which is a perfect square:

$$105^3 - 104^3 = 32761 = 181^2.$$

It is easy to see that $181 = 9^2 + 10^2$. Other instances of the general result are

$$1456^3 - 1455^3 = (35^2 + 36^2)^2$$

and

$$20273^3 - 20272^3 = (132^2 + 133^2)^2.$$

We began the chapter with the observation that the sum of the first few odd numbers is square, no matter how many are taken; we have just looked at sums and differences of cubes that are square. Let us draw the two themes together with the question of whether the sum of consecutive odd cubes is square.

In general the answer is "no." But we do have

$$1^3 + 3^3 + 5^3 + 7^3 + 9^3 = 35^2 = 5^2 \times 7^2$$
$$1^3 + 3^3 + 5^3 + 7^3 + \cdots + 57^3 = 1189^2 = 29^2 \times 41^2.$$

as well as

$$1^3 - 2^3 + 3^3 - 4^3 + 5^3 = 9^2 = 3^4$$
$$1^3 - 2^3 + 3^3 - 4^3 + \cdots - 12^3 + 13^3 = 35^2.$$

Both of these are parts of larger patterns **(13)**.

Exercises on the Notes

The exercises are keyed to the boldface numerals in the text; at the beginning of each is an indication of the background necessary and the difficulty. Unless otherwise stated, they can be done by a high school student with the appropriate prerequisites.

1. *Manipulative skill with binomial coefficients; induction. Hard.*

(a) Use the method to find fifth powers.

(b) Think of the array as consisting of triangular pieces, with the last number of each row of each triangular piece being crossed out. Consider these numbers in some detail; what do they have to do with the binomial coefficients of $(1 + t)^k$, where k is the exponent of the powers sought? Formulate a conjecture.

(c) Prove that your conjecture holds. (Look at the rth triangular piece and use induction on r.)

2. *Manipulative skill with binomial coefficients; induction. Hard.*
   (a) Let the left column of each table be the *zeroth* column. Determine the nth term in the kth column of Tables I and II.
   (b) What is the nth term in the kth column of Table III?
   (c) Formulate a conjecture and prove it by induction.

3. (a) *Simple algebra.* Prove by induction that

$$\sum_{r=1}^{n} r^3 = \left[\frac{1}{2} n(n+1) \right]^2.$$

   (b) *Moderately difficult combinatorics and algebra.* A square is divided into n^2 unit squares, like a chess board. Any two horizontal lines and any two vertical lines form a rectangle; the *breadth* of a rectangle is the smaller of its two dimensions (in the case of a square, its sidelength). There is one rectangle of breadth n, namely the square itself. It is not hard to check that there are 2^3 rectangles of breadth $n-1$. Prove that, for $3 \le k \le n$, there are k^3 rectangles of breadth $n-k+1$, and use this fact to construct an argument that

$$1^3 + 2^3 + \cdots + n^3 = \left[\frac{n(n+1)}{2} \right]^2.$$

4. *Simple manipulation and insight.* Prove the assertion.

5. *Simple manipulation and insight.* What are the diagonal terms of an $n \times n$ square array? Prove the assertions about the powers of n.

6. *College-level number theory; multiplicative functions. Moderate difficulty.*
   (a) Formulate an equation using the divisor function.
   (b) Verify that both sides of the equation are multiplicative.
   (c) Establish the result for prime powers and then deduce it in general.

7. *High school algebra.*
   (a) Find all pairs (u, v) of positive integers for which $(u+v)^2 = u^3 + v^3$.
   (b) Suppose that $(a_1 + a_2 + \cdots + a_m)^2 = a_1^3 + a_2^3 + \cdots + a_m^3$ and $(b_1 + b_2 + \cdots + b_n)^2 = b_1^3 + b_2^3 + \cdots + b_n^3$. Prove that

$$(a_1 b_1 + \cdots + a_i b_j + \cdots + a_m b_n)^2 = (a_1 b_1)^3 + \cdots + (a_i b_j)^3 + \cdots + (a_m b_n)^3.$$

(c) Prove that

$$[1 + 2 + 3 + \cdots + 2(n - k) + 2k + 2(n - k + 1) + \cdots + 2 + 2n]^2$$
$$= 1^3 + 2^3 + \cdots + [2(n - k)]^3 + [2k]^3 + [2(n - k + 1)]^3 + \cdots + 2^3 + [2k]^3,$$

i.e.,

$$\left[\sum_{i=1}^{2(n-k)} i + 2 \sum_{i=1}^{k} i + 2 \sum_{i=n-k+1}^{n} i \right]^2 = \sum_{i=1}^{2(n-k)} i^3 + 8 \sum_{i=1}^{k} i^3 + 8 \sum_{i=n-k+1}^{n} i^3.$$

8. *High school induction. Straightforward algebra.* Prove that for each positive integer n,

$$\sum_{i=1}^{n} i^5 + \sum_{i=1}^{n} i^7 = 2 \left(\sum_{i=1}^{n} i \right)^4.$$

9. *Insight; simple when you know how. Algebra for (a) and (b) is straightforward, but care is needed.*

(a) Define $g(n) = \sum\{i : 1 \le i \le n,\ i \text{ and } n \text{ are coprime}\}$ and $h(n) = \sum\{i^3 : 1 \le i \le n,\ i \text{ and } n \text{ are coprime}\}$. If p is prime and k is a positive integer, show that

$$g(p^k) = \frac{1}{2} p^{2k-1}(p - 1)$$

and

$$h(p^k) = \frac{1}{4} p^{2k}(p^{2k-1} - 1)(p - 1).$$

(b) Use (a) to show algebraically that when n is a power of a single prime, then

$$\sum_{d|n} (n/d)^3 h(d) = \left[\sum_{d|n} (n/d) g(d) \right]^2$$

where the sums are taken over all positive divisors of n.

(c) We cannot follow the strategy of item **(5)** in proving (b) for general n since neither side of the equation is multiplicative. However, find a straightforward argument by considering the set of numbers of the form $(n/d) \times z$ where d runs through the divisors of n, and z runs through all the numbers less than d and coprime with d.

10. *Insight.* Use a "pairing of divisors" argument to show that a number has an odd number of divisors if and only if it is square.

11. *Insight.* When does the nth locker get a keyturn? Under what circumstances does it get an odd number of key turns?

12. *Elementary number theory and simple algebra.* (a) Suppose that $(x+1)^3 - x^3 = y^2$. Show that $3(2x+1)^2 = (2y-1)(2y+1)$.

(b) Prove that, either $2y - 1 = a^2$ and $2y + 1 = 3b^2$ for some integers a and b, or else $2y - 1 = 3c^2$ and $2y + 1 = d^2$ for some integers c and d.

(c) Verify that the second option in (b) implies that $d^2 = 3c^2 + 2$ and argue that this option in fact cannot occur.

(d) Use the fact that $3b^2 = a^2 + 2$ to deduce that y can be expressed as the sum of two squares.

13. *Straightforward algebra.* Derive expressions for $1^3 + 3^3 + 5^3 + \cdots + (2n - 1)^3$ and $1^3 - 2^3 + 3^3 - 4^3 + \cdots - (2n)^3 + (2n+1)^3$, and determine conditions under which either is square. Find further numerical examples.

Notes

1. The table that yields the kth powers can be regarded as a succession of right triangular pieces, the $(r+1)$th piece having as its top row

$$rk + 1 \quad rk + 2 \quad rk + 3 \quad \cdots \quad rk + (k - 1) \quad rk + k = (r+1)k.$$

Consider the first triangular piece ($r = 0$). The last, and crossed-out, element in the first row is k, in the second is $\sum_{i=1}^{k-1} i = \binom{k}{2}$, and, in the mth for $3 \le m \le k$,

$$\sum_{i=1}^{k-m+1} \binom{i + m - 2}{m - 1} = \binom{k}{m}.$$

Suppose, as an induction hypothesis, that the last, and crossed out, element in the mth row of the rth triangular piece is $r^m \binom{k}{m}$, $(1 \le m \le k)$. Clearly, the first row of the $(r+1)$th triangular piece ends in $(r+1)\binom{k}{1}$. The second row of this piece is

$$r^2 \binom{k}{2} + rk + 1 \quad r^2 \binom{k}{2} + 2rk + 3 \quad \cdots \quad r^2 \binom{k}{2} + irk + \binom{i+1}{2}$$

$$\cdots \quad r^2 \binom{k}{2} + (k-1)rk + \binom{k}{2} = (r+1)^2 \binom{k}{2}.$$

The ith element of the third row is

$$r^3 \binom{k}{3} + ir^2 \binom{k}{2} + \binom{i+1}{2} r \binom{k}{1} + \binom{i+2}{3};$$

the last, and crossed-out, element for $i = k - 2$ is

$$r^3 \binom{k}{3} + r^2(k-2)\binom{k}{2} + r\binom{k-1}{2}\binom{k}{1} + \binom{k}{3}$$

$$= r^3 \binom{k}{3} + 3r^2 \binom{k}{3} + 3r\binom{k}{3} + \binom{k}{3} = (r+1)^3 \binom{k}{3}.$$

Continuing on in a similar way and using $\sum_{i=1}^{s} \binom{i+t}{t+1} = \binom{s+t+1}{t+2}$, we find that the ith element of the mth row is

$$r^m \binom{k}{m} + ir^{m-1}\binom{k}{m-1} + \binom{i+1}{2}r^{m-2}\binom{k}{m-2} + \cdots + \binom{i+m-1}{m},$$

with the last, and crossed-out, element

$$r^m \binom{k}{m} + r^{m-1}(k-m+1)\binom{k}{m-1} + r^{m-2}\binom{k-m+2}{2}\binom{k}{m-2}$$

$$+ r^{m-3}\binom{k-m+3}{3}\binom{k}{m-3} + \cdots + \binom{k}{m}.$$

Using the identity,

$$\binom{k}{b}\binom{k-b}{a} = \binom{a+b}{b}\binom{k}{a+b},$$

we find that this is

$$r^m \binom{k}{m} + mr^{m-1}\binom{k}{m} + \binom{m}{2}r^{m-2}\binom{k}{m} + \cdots + \binom{k}{m} = (r+1)^m \binom{k}{m}.$$

In particular, the sole element in the kth row is $(r+1)^k$.

The same type of process can be used to obtain factorials rather than powers:

1	2	3	4	5	6	7	8	9	10	11	12	13	14	15
	2		6	11		18	26	35		46	58	71	85	
			6			24	50			96	154	225		
						24				120				
										120				

See the articles by Calvin T. Long in *American Mathematical Monthly* 73 (1966), 846–851; *Mathematical Gazette* 66 (1982), 273–277 and *Mathematics Teacher* 75 (1982), 413–415, and by John G. Slater in *Mathematical Gazette* 67 (1983), 288–290.

2. In each of Tables I, II, III, IV, call the leftmost column the zeroth column. For $k = 1, 2, \ldots$, the nth term in the kth column of Table I is

$$\binom{n+k-2}{k-1}.$$

That each column is the running total of its predecessor follows from

$$\sum_{m=1}^{n} \binom{m+k-2}{k-1} = \sum_{m=1}^{n} \left[\binom{m+k-1}{k} - \binom{m+k-2}{k} \right]$$
$$= \binom{n+k-1}{k},$$

where we adopt the convention that $\binom{r}{s} = 0$ when $r < s$.

In Table II, the nth term of the kth column is

$$\binom{n+k-1}{k} + \binom{n+k-2}{k}$$

and once again it can be checked that each column is the running total of its predecessor. Thus the nth term of the kth column of Table III is

$$\binom{n+k-2}{k-1}\binom{n+k-1}{k} + \binom{n+k-2}{k-1}\binom{n+k-2}{k}$$
$$= \frac{1}{(k-1)!k!} \left[(n+k-2)\cdots n \right]^2 (2n+k-2)$$

(where, when $k = 1$, the term in square brackets is taken as 1). It has to be shown that, for $k \geq 1$,

$$\frac{1}{(k-1)!k!} \sum_{m=1}^{n} \left[(m+k-2)\cdots m \right]^2 (2m+k-2) = \binom{n+k-1}{k}^2$$

or

$$k \sum_{m=1}^{n} \left[(m+k-2)\cdots m \right]^2 (2m+k-2) = \left[(n+k-1)\cdots n \right]^2.$$

This is evident for $k = 1$ and any n as well as for $n = 1$ and any k.

Suppose, as an induction hypothesis, the equation holds for a fixed k and for $n = r \geq 1$. Then

$$k \sum_{m=1}^{r+1} \left[(m+k-2)\cdots m \right]^2 (2m+k-2)$$
$$= \left[(r+k-1)\cdots r \right]^2 + k \left[(r+k-1)\cdots (r+1) \right]^2 (2r+k)$$
$$- \left[(r+k-1)\cdots (r+1) \right]^2 \left[r^2 + 2rk + k^2 \right]$$
$$= \left[(r+k)(r+k-1)\cdots (r+1) \right]^2$$

i.e., when $n = r + 1$.

3. Note 1262 in *Mathematical Gazette* 21 (1937), 412–413 gives a method that is described in Additional Exercise 2. Other diagrammatic proofs appear in Classroom Notes 300 and 305 in *Mathematical Gazette* 58 (1974), 50, 211, and in *Mathematics Magazine* 58 (1985), 11; 65 (1992), 185. Exercise (b) is based on Problem 3792 in *American Mathematical Monthly* 43 (1936), 435; 45 (1938), 391. See also Notes 1263, 2687 and 2775 in *Mathematical Gazette* 21 (1937), 414–415; 41 (1957), 122; 42 (1958), 223 as well as Problem E21 in *American Mathematical Monthly* 40 (1933), 110. D. Desbrow and D. Singmaster, in separate articles in *Mathematical Gazette* 66 (1982), 97–100; 100–104 discuss ways of deriving the formulae for the sum, square sum and cube sum of the first n natural numbers. K. R. Sastry in *Mathematics and Informatics Quarterly* 6 (1996), 10–13 wonders what systematic sequences of numbers, other than the triangular ones, are such that the difference of the squares of consecutive entries are cubes.

One can investigate further examples of pairs r and s or integers for which there are exponents h and k with

$$(1^r + 2^r + \cdots + n^r)^h = (1^s + 2^s + \cdots + n^s)^k.$$

The only possibility is $(r, s; h, k) = (1, 3; 2, 1)$. See Problem E2136 in *American Mathematical Monthly* 75 (1968), 1113; 76 (1969), 946–947. Problem E2951 in *American Mathematical Monthly* 89 (1982), 424; 91 (1984), 142–143, takes up the same question for sums of odd powers. In *American Mathematical Monthly* 92 (1985), 729–731, D. E. Penney and C. Pomerance generalize the situation of powers of terms in arithmetic progression.

4. The sum of the terms in the nth row is the sum of the first $\binom{n+1}{2}$ odd numbers, less the sum of the first $\binom{n}{2}$ odd numbers, namely

$$\binom{n+1}{2}^2 - \binom{n}{2}^2 = \left[\binom{n+1}{2} - \binom{n}{2}\right]\left[\binom{n+1}{2} + \binom{n}{2}\right]$$
$$= n \cdot n^2 = n^3.$$

5. The sum of the diagonal elements is

$$1 + [2n + 3] + [4n + 5] + \cdots + [2(n-1)n + (2n-1)]$$
$$= 2n[1 + 2 + \cdots + (n-1)] + [1 + 3 + 5 + \cdots + (2n-1)]$$
$$= n^2(n-1) + n^2 = n^3.$$

6. If $\tau(n)$ represents the number of positive integers d that divide evenly into n (in symbols, $d|n$), the assertion is that

$$\sum_{d|n} \tau(d)^3 = \left[\sum_{d|n} \tau(d)\right]^2 \tag{$*$}$$

The proof of this involves a little number theory. Call a function f from the set of positive integers to itself *multiplicative* if and only if $f(mn) = f(m)f(n)$ whenever the greatest common divisor of m and n is 1. Then one requires the following propositions:

(i) for any multiplicative function f,

$$f(n) = \prod f(p^a)$$

where $n = \Pi p^a$ is the factorization of n as the product of distinct prime powers and the product $\Pi f(p^a)$ has one factor for each such prime power;

(ii) if f is multiplicative, then so also are $f(n)^r$ for each positive integer r and $\sum_{d|n} f(d)$, where the sum is taken over all positive divisors d of n.

Using (ii), we see that each side of equation $(*)$ is multiplicative, so that, by (i), it suffices to check $(*)$ when n is a prime power p^a. But this amounts to showing that $\left[1^3 + 2^3 + \cdots + (a+1)^3\right] = \left[1 + 2 + \cdots + (a+1)\right]^2$.

For more details, consult Essay Ten on page 72 of Ross Honsberger, *Ingenuity in Mathematics* (MAA). This result is also given as problem M140 in *Quantum* 5 (no. 2, March/April, 1995), 29, 57.

7.   Yin's article, in Chinese, appears in *Mathmedia, Institute of Mathematics, Academia Sinica* (Taipei, Taiwan), Volume 19, No. 2 (1995), 73–75.

There are only two pairs (u, v) of positive integers for which $(u+v)^2 = u^3 + v^3$. This equation leads to $u + v = u^2 - uv + v^2$ or $v^2 - (u+1)v + u^2 - u = 0$, which, as a quadratic in v, has discriminant $-3(u-1)^2 + 4$. This is positive only for $u = 1$ and $u = 2$. We obtain $(u, v) = (1, 2)$ and $(2, 2)$ up to order of terms.

Observe that

$$\left(\sum a_i b_j\right)^2 = \left[\left(\sum a_i\right)\left(\sum b_j\right)\right]^2 = \left(\sum a_i\right)^2\left(\sum b_j\right)^2$$
$$= \sum a_i^3 \sum b_j^3 = \sum (a_i b_j)^3.$$

It is straightforward to check that each side of

$$\left[\sum_{i=1}^{2(n-k)} i + 2\sum_{i=1}^{k} i + 2\sum_{i=n-k+1}^{n} i\right]^2 = \sum_{i=1}^{2(n-k)} i^3 + 8\sum_{i=1}^{k} i^3 + 8\sum_{i=n-k+1}^{n} i^3$$

is equal to the square of

$$(n-k)(2n-2k+1) + k(k+1) + n(n+1) - (n-k)(n-k+1)$$
$$= 2n^2 - (2k-1)n + k(2k+1).$$

8. This interesting fact was posed as Problem 419 in *American Mathematical Monthly* 21 (1914), 226. The induction step depends on the manipulation

$$2\left(\frac{n(n+1)}{2}\right)^4 + (n+1)^5 + (n+1)^7$$

$$= (n+1)^4\left[\frac{n^4}{8} + (n+1) + (n+1)^3\right]$$

$$= \frac{(n+1)^4}{8}[n^4 + 8n^3 + 24n^2 + 32n + 16]$$

$$= 2\left[\frac{(n+1)^4(n+2)^4}{16}\right].$$

9. With $g(n)$ equal to the sum of the positive numbers less than and coprime with n and $h(n)$ equal to the sum of their cubes, the assertion is that

$$\sum_{d|n}(n/d)^3 h(d) = \left[\sum_{d|n}(n/d)g(d)\right]^2.$$

The proof depends on this fact:

each number m with $1 \le m \le n$ can be written in exactly one way as a product rs, where r is a divisor of n and s is less than and coprime with n/r.

For such a number m, let r be the greatest common divisor of m and n. With this in hand, we find that the equation really is our friend $1^3 + 2^3 + \cdots + n^3 = (1 + 2 + \cdots + n)^3$.

10. All divisors of any number n, except its square root of a square number, can be paired off so that the product of the numbers of each pair is n. When n is not a perfect square this accounts for all the divisors; when n is square, its square root is an integer and a divisor without a mate.

11. This problem appeared on the 1967 Putnam Competition (*American Mathematical Monthly* 75 (1968), 735, 738). The state of the locker numbered n is changed on the kth pass if and only if k is a divisor of n. The lockers that are eventually unlocked correspond to the values of n that have an odd number of divisors; these are precisely the perfect squares.

12. Suppose that $(x+1)^3 - x^3 = y^2$. Then $3(x^2 + x) + 1 = y^2$ from which $3(2x+1)^2 = 4y^2 - 1 = (2y-1)(2y+1)$. Since $2y-1$ and $2y+1$ are relatively prime, either

(i) $2y - 1 = a^2$ and $2y + 1 = 3b^2$ for integers a and b, or

(ii) $2y - 1 = 3c^2$ and $2y + 1 = d^2$ for integers c and d.

Case (ii) can be ruled out, since it leads to $d^2 = 3c^2 + 2$, i.e., the impossibility of a square with remainder 2 upon division by 3. Hence, case (i) obtains. Thus $4y = a^2 + 3b^2 = 2(a^2 + 1)$. Since a is odd, we can write $a = 2u + 1$ and get $y = u^2 + (u + 1)^2$.

A discussion of arguments establishing this result appears in the solution to Problem 76.E in *Mathematical Gazette* 77 (1993), 126.

The same result appears as Problem E702 in *American Mathematical Monthly* 53 (1946), 36, 464–465. The *Monthly* solution solves the pellian equation $(2y)^2 - 3(2x + 1)^2 = 1$ and shows that the acceptable values of x (7, 104, 1455, 20272, ...) satisfy the recursion equation

$$x_{n+1} = 14x_n - x_{n-1} + 6$$

$$= 15x_n - 15x_{n-1} + x_{n-2}.$$

It is posed again as Problem 4299 in *American Mathematical Monthly* 55 (1948), 321; 57 (1950), 189–190.

13. The sum $\sum_{k=1}^{n}(2k - 1)^3$ of the first n odd cubes is $n^2(2n^2 - 1)$. This is square if and only if $2n^2 - 1 = m^2$ for some integer m. An equation of the form $m^2 - 2n^2 = -1$ is a Pell's equation (see Chapter 4) and has infinitely many solutions (m_r, n_r) given by

$$m_r + n_r\sqrt{2} = (1 + \sqrt{2})^{2r-1} \quad (r \in \mathbf{N}).$$

Thus, for example, $m_2 + n_2\sqrt{2} = (1 + \sqrt{2})^3 = 7 + 5\sqrt{2}$, whence $(m_2, n_2) = (7, 5)$. The next few solutions are $(41, 29)$, $(239, 169)$, $(1393, 985)$. This problem is discussed in Note 67.27 in *Mathematical Gazette* 67 (1983), 212–215.

In *American Mathematical Monthly* 7 (1900), 176, it is required that both $2n - 1$ and $2n^2 - 1$ be square. Suppose that $2n - 1 = (2r - 1)^2$. Then $n = 2r^2 - 2r + 1 = r^2 + (r - 1)^2$. Thus $2[r^2 + (r - 1)^2]^2 = s^2 + 1$ for some integer s. Noting the identity $2(p^2 + q^2)^2 = [(p+q)^2 - 2q^2]^2 + [(p-q)^2 - 2q^2]^2$, we try $p = r$, $q = r - 1$, $\pm 1 = (p - q)^2 - 2q^2 = 1 - 2(r - 1)^2$. This leads to $r = 0, 1, 2$ and $n = 1, 5$.

As for the sums with alternating signs, we have

$$1^3 - 2^3 + \cdots - (2n)^3 + (2n + 1)^3 = \sum_{k=1}^{2n+1} k^3 - 2\sum_{k=1}^{n}(2k)^3$$

$$= (n + 1)^2(4n + 1)$$

which is square whenever $n = (m^2 - 1)/4$ for odd integer m. On the other hand, note that

$$-1^3 + 2^3 - \cdots - (2n-1)^3 + (2n)^3 = 2\sum_{k=1}^{n}(2k)^3 - \sum_{k=1}^{2n}k^3$$
$$= n^2(4n+3)$$

is never square, since $4n + 3$ cannot be square.

In an article on cubes of natural numbers in arithmetic progression, appearing in *Crux Mathematicorum* 18 (1992), 161–164, K.R.S. Sastry observes that $1^3 + 3^3 + \cdots + (2n-1)^3$ is equal to $(1/2)(2n^2 - 1)(2n^2)$, a "triangular number" representing the number of dots in a equilateral triagular array with $2n^2 - 1$ dots to a side. We can more generally define *polygonal numbers* (representing the number of dots in a certain n-gonal array with r dots to a side) by the formula

$$P(n, r) = (n-2)\frac{r^2}{2} - (n-4)\frac{r}{2}.$$

Suppose that $S_k = a_1^3 + a_2^3 + \cdots + a_k^3$ is the sum of k cubes in arithmetic progression with $a_1 = 1$ and common difference d. Sastry shows that, when $d = 1$, S_k is a square number, while when $d = 2e$, then S_k is an n-gonal number of side $n = 4e+2$. When $d = 2$, S_k is both a 6-gonal and a triangular number, since $P(6, r) = P(3, 2r - 1)$ for each r.

An interesting article on another aspect of pentagonal numbers is

George E. Andrews, Euler's pentagonal number theorem, *Mathematics Magazine* 56 (1983), 279–284.

Additional Exercises

These exercises may be attempted by high school students.

1. Let n be a positive integer and let an $n \times n$ square array of numbers be formed for which the element in the ith row and the jth column ($1 \leq i, j \leq n$) is the smaller of i and j. For $n = 4$, the array would be

$$\begin{array}{cccc}
1 & 1 & 1 & 1 \\
1 & 2 & 2 & 2 \\
1 & 2 & 3 & 3 \\
1 & 2 & 3 & 4
\end{array}$$

Show that the sum of all the numbers in the array is

$$1^2 + 2^2 + 3^2 + \cdots + n^2.$$

2. Consider the infinite array of positive integers that consists of the body of the "multiplication table":

1	2	3	4	5	6	7	8	9	$\cdots$
2	4	6	8	10	12	14	16	18	$\cdots$
3	6	9	12	15	18	21	24	27	$\cdots$
4	8	12	16	20	24	28	32	36	$\cdots$
5	10	15	20	25	30	35	40	45	$\cdots$

$\cdots$

Thus, the number in the ith row and the jth column is $i \times j$.

(a) The fourth "gnomon" is indicated in the table. In general, the nth gnomon is the set of $2n - 1$ numbers consisting of the first n numbers in the nth row along with the first $n - 1$ numbers in the nth column. What is the sum of the numbers in the nth gnomon? Why?

(b) Explain why the sum of the numbers in the $n \times n$ square array in the upper left corner of the multiplication table is $(1 + 2 + 3 + \cdots + n)^2$,

(c) Use (a) and (b) to account for the formula

$$1^3 + 2^3 + 3^3 + \cdots + n^3 = (1 + 2 + 3 + \cdots + n)^2 .$$

(d) Suppose that we form a sum as follows. For each positive integer n, add together

the nth number in the first column

the $(n - 1)$th number in the third column

the $(n - 2)$th number in the fifth column

$\cdots$

the first number in the $(2n - 1)$th column.

Account for the sum.

For example, when $n = 1, 2, 3, 4$, we get

$$1 = 1^2$$

$$2 + 3 = 5 = 1^2 + 2^2$$

$$3 + 6 + 5 = 14 = 1^2 + 2^2 + 3^2$$

$$4 + 9 + 10 + 7 = 30 = 1^2 + 2^2 + 3^2 + 4^2.$$

3. Suppose that n and k are two positive integers with $k \geq 2$. Show that n^k can be written as the sum of n consecutive odd integers.

4. Find a generalization of the following set of numerical equations:

$$1 = 1$$

$$8 = 1 + 7$$

$$27 = 1 + 9 + 17$$

$$64 = 1 + 11 + 21 + 31$$

$$125 = 1 + 13 + 25 + 37 + 49.$$

5. A number is *triangular* if and only if it is of the form $\frac{1}{2}k(k+1) = 1+2+\cdots+k$. The smallest pair of consecutive positive cubes whose difference is a triangular number is $\{6^3, 5^3\}$. Indeed, $6^3 - 5^3 = 91 = \frac{1}{2} \times 13 \times 14 = 46^2 - 45^2$.

(a) Determine other pairs of consecutive cubes whose differences are triangular.

(b) Prove that the difference has the form $(z+1)^2 - z^2$, where z is a multiple of 9.

6. Let n be a positive integer. Determine the sum of the numbers in the diamond array

$$
\begin{array}{ccccccc}
 & & & 1 & & & \\
 & & 2 & & 2 & & \\
 & 3 & & 3 & & 3 & \\
 & & & \cdots & & & \\
n & & n & \cdots & & n & n \\
 & & & \cdots & & & \\
 & 2n-2 & & & 2n-2 & & \\
 & & & 2n-1 & & &
\end{array}
$$

where, for $1 \le k \le n$, row k and row $2n - k$ each have k elements, respectively equal to k and $2n - k$.

7. Make up a table of values, showing in the rows, the positive integer n and the sums $S_1 = 1 + 2 + \cdots + n$, $S_2 = 1^2 + 2^2 + \cdots + n^2$ and $S_4 = 1^4 + 2^4 + \cdots + n^4$.

(a) Express the quotients S_4/S_2 in lowest terms. What denominators do you get?

(b) Investigate possible relationships between this quotient and S_1.

(c) Using the fact that $S_1 = n(n+1)/2$ and $S_2 = n(n+1)(2n+1)/6$, conjecture and prove by induction a formula for S_4.

8. Find all values of the positive integers m and n for which

$$1! + 2! + 3! + \cdots + n! = m^2.$$

9. (a) Verify that

$$n + 2(1 + 2 + 3 + \cdots + n) = (n+1)^2 - 1$$

$$n + 3(1 + 2 + 3 + \cdots + n) + 3(1^2 + 2^2 + 3^2 + \cdots + n^2) = (n+1)^3 - 1$$

$$n+4(1+2+3+\cdots+n)+6(1^2+2^2+3^2+\cdots+n^2)+4(1^3+2^3+3^3+\cdots+n^3)$$
$$= (n+1)^4 - 1.$$

(b) Let $S_r(n) = 1^r + 2^r + \cdots + n^r = \sum_{i=1}^{n} i^r$. Prove that, for all positive integers m and n:

$$n + \binom{m}{1} S_1(n) + \binom{m}{2} S_2(n) + \cdots + \binom{m}{m-1} S_{m-1}(n) = (n+1)^m - 1.$$

10. Let n be a fixed positive integer, and let

$$u_1(n) = 1^2 + 2^2 + \cdots + n^2$$
$$u_{m+1}(n) = u_m(1) + u_m(2) + \cdots + u_m(n) \quad \text{for } m \geq 1.$$

Prove that, for each positive integer m,

$$(m + 2)! u_m(n) = n(n + 1)(n + 2) \cdots (n + m)(2n + m).$$

11. What is the largest number that cannot be written as the sum of distinct squares?

Solutions

1. *First solution.* We can visualize the result by imagining an $n \times n$ checkerboard. Begin by placing a checker on each square (n^2 checkers); place an additional checker on every square not in the first row or the first column ($(n-1)^2$ checkers); then place another checker on every square not in the first two rows or the first two columns ($(n-2)^2$ checkers). Continue on in this way to obtain an allocation of $n^2 + (n-1)^2 + (n-2)^2 + \cdots + 2^2 + 1^2$ checkers; the number of checkers placed on the square in the ith row and jth column is the smaller of i and j.

Second solution. Algebraically, we see that the sum of the numbers in the kth gnomon consisting of the numbers not exceeding k in the kth row and the kth column is $2(1 + 2 + \cdots + (k-1)) + k = k^2$. The result follows.

See Note 1262 in *Mathematical Gazette* 21 (1937), 412–413.

2. (a) The sum of the numbers in the nth gnomon is

$$2n\big(1 + 2 + \cdots + (n-1)\big) + n^2 = n^2(n-1) + n^2 = n^3.$$

(b) The sum of the numbers in the $n \times n$ upper left square array is

$$(1+2+\cdots+n)+2(1+2+\cdots+n)+3(1+2+\cdots+n)+\cdots+n(1+2+\cdots+n)$$
$$= (1 + 2 + \cdots + n)^2.$$

(d) We can show by induction that

$$n + 3(n-1) + 5(n-2) + \cdots + (2n-1) \times 1 = 1^2 + 2^2 + \cdots + n^2.$$

Assuming its truth for $n = k$, we have that

$$(k+1) + 3k + 5(k-1) + \cdots + (2k-1) \times 2 + (2k+1) \times 1$$
$$= \big[k + 3(k-1) + 5(k-2) + \cdots + (2k-1) \times 1\big]$$
$$+ \big[1 + 3 + 5 + \cdots + (2k+1)\big]$$
$$= [1^2 + 2^2 + \cdots + k^2] + (k+1)^2.$$

The result follows.

See Note 1262 in *Mathematical Gazette* 21 (1937), 412–413.

3. Consider for example 5^4. The average of five integers with this sum is 5^3, so this should be the middle terms of those summed. Indeed, $625 = 121 + 123 + 125 + 127 + 129$. In general, if $n = 2m + 1$, then

$$n^k = \sum_{i=-m}^{m} (n^{k-1} + 2i);$$

if $n = 2m$, then

$$n^k = \sum_{i=-m}^{m-1} (n^{k-1} + (2i+1)).$$

(Problem E726 in *American Mathematical Monthly* 53 (1946), 333; 54 (1947), 165–166.)

4. A generalization is

$$n^3 = \sum_{i=0}^{n-1} \big[2i(n+1) + 1\big].$$

See *Mathematics Magazine* 63 (1990), 349 for a "proof without words."

5. (a) We need to solve the equation

$$(x+1)^3 - x^3 = \frac{1}{2}y(y+1).$$

This can be rewritten in the form $v^2 - 6u^2 = 3$, where $u = 2x + 1$ and $v = 2y + 1$. Some solutions are $(v, u) = (3, 1), (27, 11), (267, 109)$. Using the theory of Pell's equation, one finds that if (v, u) is a solution, then so also is $(5v+12u, 2v+5u)$; each solution in the chain starting with $(3, 1)$ consists of odd integers. From these solutions (v, u) we can derive solutions (x, y) of the original equation. Indeed, using the relationship between (v, u) and (x, y), we can deduce that if (x, y) is a solution, then so also is $(5x+2y+3, 12x+5y+8)$. This table lists the first few solutions:

x	y	$(x+1)^3 - x^3$
0	1	$1 = 1^2 - 0^2$
5	13	$91 = 46^2 - 45^2$
54	133	$8911 = 4456^2 - 4455^2$
539	1321	$873181 = 436591^2 - 436590^2$
5340	13081	$85562821 = 42781411^2 - 42781410^2$

(b) $(x+1)^3 - x^3 = (z+1)^2 - z^2$ if and only if $2z = 3x(x+1)$. Now z is divisible by 9 if and only if either x or $x+1$ is divisible by 3. We show that if this is true of any solution in the chain of (a), then it is true of the next.

Suppose $x \equiv 0 \pmod{3}$. Since $y(y+1) = 6x^2 + 6x + 2 \equiv 2 \pmod{3}$, we must have $y \equiv 1 \pmod{3}$. Hence $5x + 2y + 3 \equiv 2 \pmod{3}$ and $(5x + 2y + 3) + 1$ is a multiple of 3. Suppose $x + 1 \equiv 0 \pmod{3}$. Then, again, $y \equiv 1 \pmod{3}$ and one deduces that $5x + 2y + 3$ is a multiple of 3. The result follows.

6. See the "brainteaser" B147 in *Quantum* 5 (no. 6, July/August, 1995), 15, 60. Observe that, for $1 \le k \le n - 1$, the sum of all the numbers in the kth and $(2n - k)$th rows together is $k[k + (2n - k)] = 2nk$, so that the sum of all the numbers is $2n \sum_{k=1}^{n-1} k + n^2 = n^3$.

7. It turns out that $5S_4/S_2 = 6S_1 - 1$ or $5S_4 + S_2 = 6S_1 S_2$, so that $5S_4 = (3n^2 + 3n - 1)S_2$. Therefore

$$S_4 = \frac{n(n+1)(2n+1)(3n^2 + 3n - 1)}{30}.$$

Problem E369 in *American Mathematical Monthly* 46 (1939), 107; 47 (1940), 112 asks when S_4 is divisible by S_2. In Q656 of *Mathematics Magazine* 52 (1979), 47, 55, the equation $5S_4 + S_2 = 6S_1 S_2$ is used to establish that either $S_1 - 1$ or S_2 is divisible by 5 for each n.

A very general result, dating back to the work of Johann Faulhaber in 1631 (!) is that $S_r = 1^r + 2^r + \cdots + n^r$ can be written as a polynomial in S_1 with coefficients independent of n multiplied, when r is odd and at least 3, by S_1^2 and, when r is even, by S_2. This is presented in a paper of A. W. F. Edwards in *American Mathematical Monthly* 93 (1986), 451–455. See also Problem 10290 in *American Mathematical Monthly* 100 (1993), 290; 102 (1995), 933–934, where several references are provided.

An elegant approach to Faulhaber's theorem is given by H. K. Krishnapriyan in his paper on Eulerian polynomials and Faulhaber's result in *College Mathematics Journal* 26 (1995), 118–123.

A. F. Beardon, in *American Mathematical Monthly* 103 (1996), 201–213 studies Faulhaber's as well as other polynomial relations among the quantities S_r. For example, we have that

$$S_1^3 = \frac{1}{4}S_3 + \frac{3}{4}S_5$$

$$S_1^5 = \frac{1}{16}S_5 + \frac{5}{8}S_7 + \frac{5}{16}S_9$$

$$8S_1^3 + S_1^2 - 9S_2^2 = 0$$

$$81S_2^4 - 18S_2^2 S_3 + S_3^2 - 64S_3^3 = 0.$$

For each pair of integers r, s with $1 \leq r < s$, there is a unique irreducible polynomial $p_{r,s}$ with integer coefficients for which $p_{r,s}(S_r, S_s) = 0$.

8. Since $r!$ ends in 0 for $r \geq 5$, it follows that $1! + 2! + \cdots + n!$ ends in 3 for $n \geq 4$, and so cannot be square. The only possibilities are $(n, m) = (1, 1)$ and $(3, 3)$. (Q657, *Mathematics Magazine* 52 (1979), 47, 55) A whimsical version of this problem appears as a "trickie" in *Mathematics Magazine* 37 (1964), 126, 82.

9. (b) Note that

$$\sum_{r=0}^{m} \binom{m}{r} S_r(n) = \sum_{r=0}^{m} \binom{m}{r} \sum_{k=1}^{n} k^r = \sum_{k=1}^{n} \sum_{r=0}^{m} \binom{m}{r} k^r$$

$$= \sum_{k=1}^{n} (1 + k)^m = S_m(n + 1) - 1.$$

Hence

$$\sum_{r=0}^{m-1} \binom{m}{r} S_r(n) = S_m(n + 1) - S_m(n) - 1$$

$$= (n + 1)^m - 1.$$

In *Mathematics Magazine* 53 (1980), 92–96, Barbara Turner uses the binomial theorem to derive a number of identities of this sort for sums of consecutive powers. Further material on the sums of the first rth powers can be found in *American Mathematical Monthly* 68 (1961), 149–151; 91 (1984), 394–403; 93 (1986), 451–455; *Mathematical Gazette* 66 (1982), 22–28; 71 (1987), 144–146; *Mathematics Magazine* 57 (1984), 296–297; 61 (1988), 189–191; 65 (1992), 38–40.

10. This is generalized in Problem 4380, *American Mathematical Monthly* 57 (1950), 119; 58 (1951), 429. The result clearly holds for $m = 1$ and all n as well as for $n = 1$ and all m. Suppose that it is known for $m = k$

and all n and for $m = k + 1$ and $1 \le n \le r$. Then

$$(k+3)!u_{k+1}(r+1)$$

$$= (k+3)\big[(k+2)!u_k(r+1)\big] + (k+3)!u_{k+1}(r)$$

$$= (k+3)\big[(r+1)(r+2)\cdots(r+1+k)(2r+2+k)\big]$$

$$\quad + r(r+1)\cdots(r+1+k)(2r+1+k)$$

$$= \big[(r+1)(r+2)\cdots(r+1+k)\big]\big[k^2 + (2r+5)k + (6r+6) + rk + 2r^2 + r\big]$$

$$= \big[(r+1)(r+2)\cdots(r+1+k)\big]\big[(2r^2+7r+6) + (3r+5)k + k^2\big]$$

$$= (r+1)(r+2)\cdots(r+1+k)(r+2+k)\big[2(r+1)+1+k\big].$$

The result follows by induction.

In *Mathematics and Informatics Quarterly* 6 (1996), 45, is quoted a theorem attributed to von Staudt. If the positive integers a and b have greatest common divisor 1, then, for each natural number m,

$$\frac{S_m(ab)}{ab} - \frac{S_m(a)}{a} - \frac{S_m(b)}{b}$$

is an integer.

11. Problem 533 in *Mathematics Magazine* 36 (1963), 319; 37 (1964), 201–202. The answer is 128. The numbers from 129 to 192 can be written as the sum of distinct squares none exceeding 10^2. Since $1^2 + 2^2 + \cdots + 10^2 = 385$, representations of numbers from 193 to 256, inclusive, can be found by subtracting from this equation representations of numbers from 129 to 192 inclusive. We now can obtain every number from 129 to 256 inclusive using distinct squares not exceeding 10^2. Adding 11^2 to these representations covers the numbers from 250 to 377 inclusive. We can continue on, adding to existing representations, squares of 12, 13, and so on, to cover all higher numbers.

CHAPTER 2

Pythagorean Triples and Their Relatives

Squares

Most of us were introduced to pythagorean triples at school. They are sets of three whole numbers for which the sum of the squares of the two smallest is equal to the square of the largest. Some common examples of such triples are $(3, 4, 5)$, $(5, 12, 13)$, $(8, 15, 17)$. Indeed, $3^2 + 4^2 = 5^2$; $5^2 + 12^2 = 13^2$; $8^2 + 15^2 = 17^2$. Many such triples were known to the Babylonians **(1)**. Early mathematicians knew that a triangle with integer sides was right exactly when its side lengths formed a pythagorean triple. As you can imagine, the study of these triples has a long and interesting history and we can here provide only a sample of their interesting properties.

Here is an algorithm for producing an infinite family of pythagorean triples:

Take any odd positive integer; this will be the smallest number.
Square the integer and express the square as the sum of two con-
secutive integers; these will be the largest two numbers.

For example, start with the odd number 31. Its square is 961, which is the sum of the consecutive integers 480 and 481. The pythagorean triple is $(31, 480, 481)$. A convenient way to check that this works is to note that we can factor a difference of squares to obtain

$$481^2 - 480^2 = (481 - 480)(481 + 480) = 1 \times 961 = 31^2.$$

The largest two entries in each of these triples differ by only 1. Because there is no square which is exactly twice another, there are no pythagorean triples whose smallest numbers are the same. However, we can concoct as many as we want where the smallest numbers differ by 1. We use the fact that if (x, y, z) is a pythagorean triple (i.e., $x^2 + y^2 = z^2$), then $(2x + y + 2z, x + 2y + 2z, 2x + 2y + 3z)$ is also a pythagorean triple whose smallest numbers have the same difference. Beginning with $(3, 4, 5)$, we obtain in succession, $(20, 21, 29)$, $(119, 120, 169)$, $(696, 697, 985)$, ...

There are many other transformations that will produce new triples from old. For example, given a triple (x, y, z), we can obtain $(x + 4y + 4z, 4x + 7y + 8z, 4x + 8y + 9z)$.

There is a formula which will generate all such triples; (x, y, z) is a pythagorean triple if and only if we can write x, y, z in the form

$$x = k(m^2 - n^2), \qquad y = 2kmn, \qquad z = k(m^2 + n^2)$$

where k, m, n are integers. The numbers x, y, z will have greatest common divisor equal to 1 if and only if we can arrange for k to equal 1 and for m and n to have greatest common divisor 1 **(2)**.

For any pythagorean triple, (x, y, z), it turns out that $z^2 + xy$ and $z^2 - xy$ can always be written as the sum of two integer squares. For example, for the triple $(12, 35, 37)$, we have the numerical equations $37^2 + 420 = 5^2 + 42^2$ and $37^2 - 420 = 7^2 + 30^2$. In general,

$$z^2 + xy = \left(\frac{x + y + z}{2}\right)^2 + \left(\frac{x + y - z}{2}\right)^2$$

and

$$z^2 - xy = \left(\frac{x - y + z}{2}\right)^2 + \left(\frac{x - y - z}{2}\right)^2.$$

Despite appearances, the terms in parentheses are all integers because it is not possible for only one or all three terms of a pythagorean triple to be odd.

The smallest two numbers of any pythagorean triple (x, y, z) can be written in the form $x = a+b$ and $y = b+c$, where $a^2 + b^2 + c^2$ is a square. For example, the triple $(x, y, z) = (8, 15, 17)$ corresponds to $(a, b, c) = (2, 6, 9)$ **(3)**.

Is it possible for two distinct pythagorean triples to have the same sum? The answer is *yes,* but such pairs are not as common as one might expect. The smallest common sum for two sets of pythagorean triples is $240 = 40 + 96 + 104 = 15 + 112 + 113$, while 1680 is the common sum of three pythagorean triples. If we require that the triples both be primitive, then the smallest sum is $1716 = 364 + 627 + 725 = 748 + 195 + 773$, while 14280 is the smallest number that is the sum of three pythagorean triples **(4)**.

To illustrate the sort of problem in which pythagorean triples are pertinent, consider the problem of finding pairs of triangles whose sides are of integer length and which have equal areas. For example, the triangle with sidelengths 5, 5, 6 has the same area as that with sidelengths 5, 5, 8. Similarly, that with lengths 13, 13, 10 has the same area as that with lengths 13, 13, 24. We can find any number of such examples. Given a pythagorean

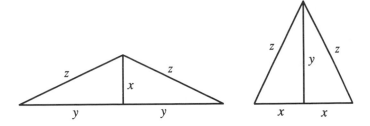

FIGURE 2.1

triple (x, y, z), we find that the isoceles triangles with side lengths z, z, $2x$ and with lengths z, z, $2y$ have the same area. The neatest way to see this is by observing that each triangle can be composed of two x-y-z right triangles as in Figure 2.1.

Pythagorean triples constitute a set with an algebraic structure, so that they can be combined with each other to produce other triples **(5)**.

Other equations involving squares

There are various equations which generalize that of Pythagoras. For example, $3^2 + 4^2 = 5^2$, is generalized by the sequence of equations:

$$3^2 + 4^2 = 5^2$$

$$10^2 + 11^2 + 12^2 = 13^2 + 14^2$$

$$21^2 + 22^2 + 23^2 + 24^2 = 25^2 + 26^2 + 27^2$$

$$36^2 + 37^2 + 38^2 + 39^2 + 40^2 = 41^2 + 42^2 + 43^2 + 44^2.$$

For each odd number, we can form a corresponding equation for which the total number of terms on both sides of the equation is that number and the two consecutive numbers whose squares appear on either side of the equals sign add up to the square of that number **(6)**.

In passing, we note the similarity between the foregoing set of equations and the following two sets **(7)**:

$$1 + 2 = 3 = 3 \times (1^2)$$

$$4 + 5 + 6 = 7 + 8 = 3 \times (1^2 + 2^2)$$

$$9 + 10 + 11 + 12 = 13 + 14 + 15 = 3 \times (1^2 + 2^2 + 3^2)$$

$$16 + 17 + 18 + 19 + 20 = 21 + 22 + 23 + 24 = 3 \times (1^2 + 2^2 + 3^2 + 4^2),$$

and

$$1 = 0^3 + 1^3$$

$$2 + 3 + 4 = 1^3 + 2^3$$

$$5 + 6 + 7 + 8 + 9 = 2^3 + 3^3$$

$$10 + 11 + 12 + 13 + 14 + 15 + 16 = 3^3 + 4^3.$$

We can also generalize the pythagorean equation by looking for squares which are the sums of three or more squares. Here are some interesting examples **(8)**:

$$0^2 + 1^2 + 0^2 = 1^2$$

$$1^2 + 2^2 + 2^2 = 3^2$$

$$2^2 + 3^2 + 6^2 = 7^2$$

$$3^2 + 4^2 + 12^2 = 13^2$$

$$4^2 + 5^2 + 20^2 = 21^2$$

In fact, 21^2 has a lot more representations **(9)**:

$$21^2 = 4^2 + 8^2 + 19^2 = 4^2 + 13^2 + 16^2 = 8^2 + 11^2 + 16^2$$

$$= 6^2 + 9^2 + 18^2 = 7^2 + 14^2 + 14^2.$$

There are infinitely many squares expressible as the sum of three other squares. Solutions to the equation $t^2 = x^2 + y^2 + z^2$ are given by

$$t = m^2 + n^2 + p^2 + q^2$$

$$x = m^2 - n^2 - p^2 + q^2$$

$$y = 2mn - 2pq$$

$$z = 2mp + 2nq \qquad \textbf{(10)}.$$

In the eighteenth century **(11)**, the Japanese mathematician Matsunago reported that $u^4 = x^2 + y^2 + z^2$ is satisfied by

$$u = m^2 + n^2 \qquad x = m^4 - n^4$$
$$y = 4m^2n^2 \qquad z = 2(m^2 - n^2)mn.$$

A magic square given by Andrew Bremner is the vehicle for expressing 65^2 as the sum of three squares in six different ways (as sums of rows and columns of squares):

15^2	20^2	60^2
36^2	48^2	25^2
52^2	39^2	0^2

For squares as sums of four squares, we have **(12)**

$$0^2 + 1^2 + 2^2 + 2^2 = 3^2$$

$$2^2 + 3^2 + 4^2 + 14^2 = 15^2$$

$$4^2 + 5^2 + 6^2 + 38^2 = 39^2$$

$$6^2 + 7^2 + 8^2 + 74^2 = 75^2$$

$$8^2 + 9^2 + 10^2 + 122^2 = 123^2$$

The square $93025 = 305^2$ can be written as the sum of four squares in at least ten different ways; it is the sum of the numbers in each of the rows, columns and diagonals of the following "magic square of squares" found by Allan William Johnson **(13)**:

900	60516	29584	2025
8649	13456	4356	66564
15876	19044	56169	1936
67600	9	2916	22500

that is,

30^2	246^2	172^2	45^2
93^2	116^2	66^2	258^2
126^2	138^2	237^2	44^2
260^2	3^2	54^2	150^2

Pythagorean triples correspond to the lengths of sides of right angled triangles. A nice three-dimensional analogue of this is the tetrahedron with side lengths 44, 117, 240, 125, 244, 267. The base of the tetrahedron is a triangle with sides 125, 244, 267. The remaining faces are three right angled triangles corresponding to the pythagorean triples $(44, 240, 244)$, $(44, 117, 125)$, $(117, 240, 267)$. Furthermore, the sum of the squares of the lengths of each pair of opposite sides is equal to the sum for any other pair:

$$117^2 + 244^2 = 125^2 + 240^2 = 44^2 + 267^2 = 73225.$$

Sums of consecutive squares which are squares or cubes

In the first chapter, we saw that, for any positive integer n, the sum of the first n cubes is a square. Is the sum of the first n squares ever a square? Of course, this is true for $n = 1$. But are there any other instances?

A high school formula for the sum of the first n squares is

$$1^2 + 2^2 + \cdots + n^2 = \frac{n(n+1)(2n+1)}{6}.$$

If this is not familiar to you, you might want to check it for several values of n. The question which we asked now becomes to find a value of n for which $[n(n + 1)(2n + 1)]/6$ is a square. A little playing around might lead you to the case $n = 24$. For, then, $n + 1 = 25, 2n + 1 = 49$ and $(n/6) = 4$, so that the quantity is the product of three squares. Thus

$$1^2 + 2^2 + \cdots + 24^2 = 70^2.$$

Can you find other values of n? The equation

$$1^2 + 2^2 + 3^2 + 4^2 + 5^2 + 6^2 + 7^2 + 8^2 + 9^2 + 10^2 + 12^2 = 23^2$$

comes close **(14)**.

There are infinitely many positive integers r for which the sum of r consecutive squares (not necessarily the first r squares) is a perfect square **(15)**. The following table gives the first few values of r for which this is true and the value of the smallest number x such that the sum of the r consecutive squares beginning with x^2 is a square.

r	1	2	11	23	24	24	26	33	47	49	50	59
x	1	3	18	7	1	25	25	7	539	25	7	22

For example, corresponding to $r = 11$, we have the equation

$$18^2 + 19^2 + 20^2 + 21^2 + 22^2 + 23^2 + 24^2 + 25^2 + 26^2 + 27^2 + 28^2 = 77^2.$$

Observe from the table that the sum of the first 24 and the second 24 squares are each square. This means that

$$\left[1^2 + 2^2 + \cdots + n^2\right]\left[(n + 1)^2 + (n + 2)^2 + \cdots + (2n)^2\right]$$

is square when $n = 24$. There are infinitely many values of n for which this quantity is square, including $n = 1, 7, 24, 120, 391, 1921$ **(16)**. From $25^2 + \cdots + 48^2 = 182^2$ and $25^2 + \cdots + 50^2 = 195^2$, we obtain $182^2 + 49^2 + 50^2 = 195^2$.

In somewhat the same vein, there are infinitely many integers m such that m^3 is equal to the sum of m consecutive squares **(17)**. Here are the smallest examples:

$$47^3 = 22^2 + 23^2 + 24^2 + \cdots + 67^2 + 68^2$$

and

$$2161^3 = 989^2 + 990^2 + \cdots + 3148^2 + 3149^2.$$

At the beginning of this section, we recalled that the sum of the first n cubes is a square. How many other examples are there of consecutive cubes adding up to a square? For example, $23^3 + 24^3 + 25^3 = 204^2$ **(18)**.

Cubes

The abundance of pythagorean triples leads us to wonder whether there are cubic versions of pythagorean triples, i.e., whether there are cubes which are the sum of two other cubes. The Swiss mathematician Leonhard Euler (1707–1783) proved over two hundred years ago that the answer is *no* **(19)**. However there are some near misses:

$$9^3 + 10^3 = 1729 = 12^3 + 1; \quad 6^3 + 8^3 = 728 = 9^3 - 1.$$

The Indian mathematician, Srinivasa Ramanujan (1887–1920), found a method of producing infinitely many cubes which can be expressed—almost—as a sum of two other cubes, up to a difference of 1 either way **(20)**. The method depends on being able to express certain expressions in the variable of x as series in ascending positive and negative powers of x, something like the binomial expansions you may have learned in high school. (If this is not familiar to you, just skip over this passage; it will not be needed for what comes later on.)

Suppose that

$$\frac{1 + 53x + 9x^2}{1 - 82x - 82x^2 + x^3} = a_0 + a_1 x + a_2 x^2 + \cdots + a_n x^n + \cdots$$
$$= u_1/x + u_2/x^2 + \cdots + u_n/x^n + \cdots;$$

$$\frac{2 - 26x - 12x^2}{1 - 82x - 82x^2 + x^3} = b_0 + b_1 x + b_2 x^2 + \cdots + b_n x^n + \cdots$$
$$= v_1/x + v_2/x^2 + \cdots + v_n/x^n + \cdots;$$

$$\frac{2 + 8x - 10x^2}{1 - 82x - 82x^2 + x^3} = c_0 + c_1 x + c_2 x^2 + \cdots + c_n x^n + \cdots$$
$$= w_1/x + w_2/x^2 + \cdots + w_n/x^n + \cdots.$$

Then it turns out that, for each integer n,

$$a_n^3 + b_n^3 = c_n^3 + (-1)^n \quad \text{and} \quad u_n^3 + v_n^3 = w_n^3 + (-1)^{n-1}.$$

Some numerical equations which are special cases of these are:

$$1^3 + 2^3 = 2^3 + 1 \qquad\qquad 9^3 + 10^3 = 12^3 + 1$$
$$135^3 + 138^3 = 172^3 - 1 \qquad 791^3 + 812^3 = 1010^3 - 1$$
$$11161^3 + 11468^3 = 14258^3 + 1 \quad 65601^3 + 67402^3 = 83802^3 + 1$$

Ramanujan once observed to G. H. Hardy that 1729 is the smallest natural number expressible as the sum of two cubes in two different ways: $1^3 + 12^3 = 9^3 + 10^3 = 1729$. Some other numbers expressible as the sum of two cubes in two ways are:

$$2^3 + 16^3 = 9^3 + 15^3$$

$$2^3 + 34^3 = 15^3 + 33^3$$

$$9^3 + 34^3 = 16^3 + 33^3$$

$$10^3 + 27^3 = 19^3 + 24^3.$$

Two consecutive numbers each expressible as the sum of two cubes are $4940 = 3^3 + 17^3$ and $4941 = 13^3 + 14^3$.

In an entertaining paper, Joseph H. Silverman shows that given any integer N, there is a positive integer A for which the equation $x^3 + y^3 = A$ has at least N solutions in integers **(21)**. He gives a number of examples.

Three representations:

$$4104 = 2^3 + 16^3 = 9^3 + 15^3 = (-12)^3 + 18^3;$$

$$3242197 = 76^3 + 141^3 = 85^3 + 138^3 = (-171)^3 + 202^3;$$

$$87539319 = 167^3 + 436^3 = 228^3 + 423^3 = 255^3 + 414^3;$$

$$175959000 = 525^3 + 315^3 = 552^3 + 198^3 = 560^3 + 70^3;$$

$$15170835645 = 517^3 + 2468^3 = 709^3 + 2456^3 = 1733^3 + 2152^3.$$

Four representations:

$$42549416 = 74^3 + 348^3 = 272^3 + 282^3$$
$$= (-2662)^3 + 2664^3 = (-475)^3 + 531^3;$$

$$26059452841000 = 4170^3 + 29620^3 = 12900^3 + 28810^3$$
$$= 14577^3 + 28423^3 = 21930^3 + 24940^3;$$

$$6963472309248 = 2421^3 + 19083^3 = 5436^3 + 18948^3$$
$$= 10200^3 + 18072^3 = 13322^3 + 16630^3.$$

Five representations:

$$1148834232 = 222^3 + 1044^3 = 718^3 + 920^3 = 816^3 + 846^3$$
$$= (-7986)^3 + 7992^3 = (-1425)^3 + 1593^3.$$

Failing to find cubes which are the sums of two other cubes, we might try to see whether there are cubes which are the sums of three other cubes. We already know the answer to this question. Since both 1 and -1 are cubes, we can rewrite a couple of our numerical equations as

$$(-1)^3 + 9^3 + 10^3 = 12^3 \quad \text{and} \quad 1^3 + 6^3 + 8^3 = 9^3.$$

An intriguing equation is $3^3 + 4^3 + 5^3 = 6^3$, since it is reminiscent of the pythagorean equation $3^2 + 4^2 = 5^2$. It is natural to conjecture that $3^4 + 4^4 + 5^4 + 6^4$ is equal to 7^4, but, alas, this fails to be true **(22)**.

There are many other examples, some involving quite small numbers, and the reader equipped with a pocket calculator might attempt to find them. Unlike the case of pythagorean triples, there does not seem to be an all-purpose formula which will give a complete set of solutions in integers to the equation

$$x^3 + y^3 + z^3 = w^3.$$

Some examples of this are

$$20^3 = 7^3 + 14^3 + 17^3 \quad \text{and} \quad 29^3 = 11^3 + 15^3 + 27^3.$$

However, there are a number of formulae which give infinite families of solutions. For example, if k is any integer, we can get solutions of the above equation from any one of the following:

  I. $x = k^3 + 1$; $y = 2k^3 - 1$; $z = k(k^3 - 2)$; $w = k(k^3 + 1)$

  II. $x = 3k^2$; $y = 6k^2 - 3k + 1$; $z = 3k(3k^2 - 2k + 1) - 1$; $w = z + 1$

  III. $x = 3k^2$; $y = 6k^2 + 3k + 1$; $z = 3k(3k^2 + 2k + 1)$; $w = z + 1$

    Two-parameter families of solutions are given by

  IV. $x = 3u^2 + 5uv - 5v^2$; $y = 4u^2 - 4uv + 6v^2$;

     $z = 5u^2 - 5uv - 3v^2$; $w = 6u^2 - 4uv + 4v^2$

  V. $x = 3u^2 + 16uv - 7v^2$; $y = 6u^2 - 4uv + 14v^2$;

    $z = -3u^2 + 16uv + 7v^2$; $w = 6u^2 + 4uv + 14v^2$.

  VI. $x = u(u^3 - v^3)$; $y = v(u^3 - v^3)$;

     $z = v(2u^3 + v^3)$; $w = u(u^3 + 2v^3)$.

  VII. $x = v(u^3 + v^3)$; $y = v(2u^3 - v^3)$;

     $z = u(u^3 - 2v^3)$; $w = u(u^3 + v^3)$.

  VIII. $x = u^7 - 3(v + 1)u^4 + (3v^2 + 6v + 2)u$;

    $y = 2u^6 - 3(2v + 1)u^3 + (3v^2 + 3v + 1)$;

    $z = u^6 - (3v^2 + 3v + 1)$; $w = u^7 - 3vu^4 + (3v^2 - 1)u$ **(23)**.

Some cubes can be written as the sum of three others in more than one way **(24)**:

$$87^3 = 26^3 + 55^3 + 78^3 = 20^3 + 54^3 + 79^3 = 38^3 + 48^3 + 79^3.$$

$$108^3 = 12^3 + 72^3 + 96^3 = 13^3 + 51^3 + 104^3 = 15^3 + 82^3 + 89^3$$

$$= 24^3 + 38^3 + 106^3 = 54^3 + 72^3 + 90^3.$$

$$870^3 = 17^3 + 687^3 + 694^3 = 537^3 + 564^3 + 687^3 = 235^3 + 485^3 + 810^3$$
$$= 200^3 + 540^3 + 790^3 = 260^3 + 550^3 + 780^3 = 380^3 + 480^3 + 790^3$$
$$= 225^3 + 630^3 + 735^3 = 330^3 + 450^3 + 810^3$$
$$= 435^3 + 580^3 + 725^3.$$

There are cubes that are the sums of consecutive cubes:

$$6^3 = 3^3 + 4^3 + 5^3$$
$$20^3 = 11^3 + 12^3 + 13^3 + 14^3$$
$$180^3 = 64 \times 45^3$$
$$= 6^3 + 7^3 + 8^3 + \cdots + 69^3 \quad \text{(64 terms)}$$
$$540^3 = 125 \times 108^3$$
$$= 34^3 + 35^3 + 36^3 + \cdots + 158^3 \quad \text{(125 terms)}$$
$$16830^3 = 1000 \times 1683^3$$
$$= 1134^3 + 1135^3 + 1136^3 + \cdots + 2133^3 \quad \text{(1000 terms)}$$

Each of the last three equations indicates a set of consecutive integers the average of whose cubes is the cube of an integer.

Cubes of other arithmetic progressions can also add up to a cube **(25)**:

$$408^3 = 149^3 + 256^3 + 363^3$$
$$440^3 = 230^3 + 243^3 + 256^3 + 269^3 + 282^3$$
$$495^3 = 15^3 + 52^3 + 89^3 + 126^3 + 163^3 + 200^3 + 237^3$$
$$+ 274^3 + 311^3 + 348^3$$
$$1155^3 = 435^3 + 506^3 + 577^3 + 648^3 + 719^3 + 790^3.$$

We finish this section with the observations:

$$13^3 = 10^3 + 9^3 + 7^3 + 5^3 = 12^3 + 7^3 + 5^3 + 1^3$$
$$14^3 = 10^3 + 9^3 + 8^3 + 7^3 + 5^3 + 3^3 + 2^3.$$

Fourth and higher powers

Based on the experience with squares and cubes, Leonhard Euler conjectured that no fourth power could be written as the sum of fewer than four fourth powers, no fifth power could be written as the sum of fewer than five fifth powers, and so on. There are fourth powers which can be written as the sum of exactly four other nonzero fourth powers **(26)**:

$$353^4 = 30^4 + 120^4 + 272^4 + 315^4$$
$$651^4 = 240^4 + 340^4 + 430^4 + 599^4.$$

The first of these equations was found by Norrie in 1911. As the identity

$$(4x^4 + y^4)^4 = (4x^4 - y^4)^4 + 2(4x^3y)^4 + 2(2xy^3)^4$$

indicates, there are infinitely many fourth powers which are the sums of five nonzero fourth powers. Likewise, there are infinitely many fourth powers whose doubles are the sum of three nonzero fourth powers:

$$2(x^2 + xy + y^2)^4 = (x^2 - y^2)^4 + (2xy + y^2)^4 + (2xy + x^2)^4.$$

Two of the more interesting examples of fourth powers expressible as the sum of five fourth powers are:

$$65^4 = 64^4 + 32^4 + 12^4 + 8^4 + 1^4$$
$$15^4 = 14^4 + 9^4 + 8^4 + 6^4 + 4^4$$

There are examples with quite small numbers in which a fifth power is expressible as the sum of five or six other fifth powers **(27)**:

$$12^5 = 11^5 + 9^5 + 7^5 + 6^5 + 5^5 + 4^5$$
$$30^5 = 29^5 + 19^5 + 16^5 + 11^5 + 10^5 + 5^5$$
$$32^5 = 28^5 + 24^5 + 22^5 + 17^5 + 16^5 + 15^5$$
$$67^5 = 66^5 + 36^5 + 31^5 + 23^5 + 18^5 + 13^5$$
$$= 66^5 + 34^5 + 31^5 + 29^5 + 20^5 + 7^5$$
$$72^5 = 67^5 + 47^5 + 46^5 + 43^5 + 19^5.$$

Note also the near miss

$$22^5 + 1^5 = 21^5 + 16^5 + 7^5 + 5^5 + 4^5.$$

However, it is only recently that Euler's conjecture has been disproved for both fourth and fifth powers. In 1966, L. J. Lander and T. R. Parkin discovered the relation

$$144^5 = 27^5 + 84^5 + 110^5 + 133^5$$

which expresses a fifth power as the sum of *four* fifth powers. It took until 1987 to find fourth powers which were the sum of only *three* other fourth powers. N. Elkies **(28)** showed that there are infinitely many such fourth powers, including

$$20615673^4 = 2682440^4 + 15365639^4 + 18796760^4$$

while R. Frye discovered

$$422481^4 = 95800^4 + 217519^4 + 414560^4.$$

The latter equation is believed to involve the smallest numbers possible.

It is much easier to find three fourth powers which sum to a square. If $x^2 + y^2 = z^2$, then the fourth powers of yz, zx, xy add up to the square of $z^4 - x^2y^2$. For example, starting with the equation $3^2 + 4^2 = 5^2$, we arrive at $12^4 + 15^4 + 20^4 = 481^2$. Not every such equation arises in this way. One which does not is

$$60^4 + 135^4 + 148^4 = 28721^2 .$$

There are numbers which can be expressed as the sum of two fourth powers in two different ways. For example, we have

$$59^4 + 158^4 = 133^4 + 134^4.$$

Other numbers can be expressed as the sum of three fourth powers in two different ways:

$$5^4 + 6^4 + 11^4 = 1^4 + 9^4 + 10^4$$
$$3^4 + 7^4 + 8^4 = 1^4 + 2^4 + 9^4$$
$$8^4 + 9^4 + 17^4 = 3^4 + 13^4 + 16^4.$$

We also have:

$$7^4 + 28^4 = 3^4 + 20^4 + 26^4$$
$$51^4 + 76^4 = 5^4 + 42^4 + 78^4.$$

There seems to be no known number expressible as the sum of two fifth powers in two different ways, but we have **(29)**

$$3^5 + 54^5 + 62^5 = 24^5 + 28^5 + 67^5$$
$$8^5 + 62^5 + 68^5 = 21^5 + 43^5 + 74^5$$
$$13^5 + 51^5 + 64^5 = 18^5 + 44^5 + 66^5$$
$$39^5 + 92^5 + 100^5 = 49^5 + 75^5 + 107^5$$

and

$$3^6 + 19^6 + 22^6 = 10^6 + 15^6 + 23^6$$

$$3^6 + 55^6 + 80^6 = 32^6 + 43^6 + 81^6$$

$$1^6 + 500^6 + 515^6 = 197^6 + 409^6 + 556^6$$

$$36^6 + 37^6 + 67^6 = 15^6 + 52^6 + 65^6$$

$$11^6 + 65^6 + 78^6 = 37^6 + 50^6 + 81^6.$$

Mixed exponents

Pythagorean triples correspond to the expression of one square as the sum of two others. We can ask about the representation of higher powers of numbers as the sum of two other powers. The most notorious question of this sort is the Fermat conjecture that there are no positive integers x, y, z for which $x^n + y^n = z^n$ when n is a positive integer exceeding 2. At this point, Andrew Wiles almost certainly has proved that this conjecture is valid **(30)**.

If we do not require all the exponents to be equal, then we can have many possibilities for equations of the form $x^p + y^q = z^r$, including

$$10^2 + 3^5 = 7^3$$

$$18^3 + 3^6 = 3^8$$

$$28^2 + 2^9 = 6^4$$

$$648^2 \mid 108^3 = 6^8$$

$$110592^2 + 4608^3 = 24^8.$$

The last four of these equations yields solutions to $x^2 + y^3 = z^4$, namely $(x, y, z) = (27, 18, 9)$, $(28, 8, 6)$, $(648, 108, 36)$, $(110592, 4608, 576)$. An additional solution is $(x, y, z) = (433, 143, 42)$ **(31)**.

The situation of special interest is when the exponents are relatively large. I recently heard a talk by A. Granville which drew attention to the Fermat–Catalan Conjecture:

> *There are only finitely many sextuples of positive integers* $(x, y, z; p, q, r)$ *for which the greatest common divisor of x, y, z is* 1, $\frac{1}{p} + \frac{1}{q} + \frac{1}{r} < 1$ *and $x^p + y^q = z^r$.*

The sextuple $(3, 10, 7; 5, 2, 3)$, for example, fails to satisfy all of the conditions we desire. So far, only ten sextuples seem to be known, yielding the equations:

$$1^n + 2^3 = 3^2 \quad \text{with } n \geq 7$$

$$2^5 + 7^2 = 3^4$$

$$7^3 + 13^2 = 2^9$$

$$2^7 + 17^3 = 71^2$$

$$3^5 + 11^4 = 122^2$$

$$17^7 + 76271^3 = 21063928^2$$

$$33^8 + 1549034^2 = 15613^3$$

$$43^8 + 96222^3 = 30042907^2$$

$$1414^3 + 2213459^2 = 65^7$$

$$9262^3 + 15312283^2 = 113^7.$$

Exercises on the Notes

2. *Algebra and number theory. Moderate.* (a) A pythagorean triple is *primitive* if and only if the greatest common divisor of its three entries is 1. Show that every pythagorean triple has the form (kx, ky, kz) where k is a positive integer and (x, y, z) is primitive.

(b) Show that for any primitive pythagorean triple (x, y, z), z and exactly one of x and y are odd.

(c) Suppose that x and z are odd. Argue that $z - x$ and $z + x$ have greatest common divisor 2. By considering $y^2 = (z - x)(z + x)$, prove that, for some integers m and n, $z - x = 2n^2$ and $z + x = 2m^2$.

(d) Obtain the general formula for pythagorean triples given in the text.

(e) Verify that, indeed, for each pair (m, n), $(m^2 - n^2, 2mn, m^2 + n^2)$ is a pythagorean triple.

3. *Insight. Algebra. Moderate.* (a) Note that we require $c - a = y - x$. Determine the triples (a, b, c) that correspond to $(x, y, z) = (3, 4, 5)$, $(20, 21, 29)$, $(5, 12, 13)$. What do you observe about $a + b + c$?

(b) Conjecture what $a, b,$ and c might be in terms of x, y, and z, and check your guess.

6, 7, 8. *Pattern recognition; simple algebra.* Conjecture and verify an algebraic identity suggested by the numerical equations.

10. *Simple algebra.* Verify that $t^2 = x^2 + y^2 + w^2$.

16. *Moderate algebra; surds.* (a) Find three distinct solutions in nonnegative pairs of integers (r, m) for $r^2 - 7m^2 = 9$ such that $1 \le r \le 20$.

(b) Verify that $(x, y) = (8, 3)$ satisfies $x^2 - 7y^2 = 1$.

(c) Define the integers u_k and v_k by

$$u_k + v_k\sqrt{7} = (8 + 3\sqrt{7})^k$$

for $k = 1, 2, \ldots$. Determine $(u_1, v_1), (u_2, v_2), (u_3, v_3)$ and show that $(x, y) = (u_k, v_k)$ satisfies $x^2 - 7y^2 = 1$. [HINT: Show that $u_k - v_k\sqrt{7} = (8 - 3\sqrt{7})^k$ and use the fact that $u_k^2 - 7v_k^2 = (u_k + v_k\sqrt{7})(u_k - v_k\sqrt{7})$.]

(e) Suppose $r^2 - 7m^2 = 9$ and $u^2 - 7v^2 = 1$. Let x and y be such that

$$x + y\sqrt{7} = (r + m\sqrt{7})(u + v\sqrt{7}).$$

What is $x^2 - 7y^2$?

(f) Show how to determine infinitely many integer solutions of $x^2 - 7y^2 = 9$. Give some of them numerically.

(g) Prove that

$$[1^2 + \cdots + n^2][(n+1)^2 + \cdots + (2n)^2] = (1/36)n^2(2n+1)^2(n+1)(7n+1)$$

and deduce that this is square if and only if $(n+1)(7n+1) = m^2$ for some integer m. Manipulate this to get the equivalent equation $(7n+4)^2 - 7m^2 = 9$, and thus indicate how to find infinitely many possibiities.

20. *Moderate algebra.* Using a pocket calculator and the factorization of each side as a sum of cubes, determine the prime factors of and prove the inequality of each side of the numerical equations

$$11161^3 + 11468^3 = 14258^3 + 1 \qquad 65601^3 + 67402^3 = 83802^3 + 1.$$

23. *High school algebra.* Suppose that $a^3 + b^3 + c^3 = d^3$ and $p^3 + q^3 + r^3 = s^3$. Consider the equation

$$(a + tp)^3 + (b + tq)^3 + (c + tr)^3 = (d + ts)^3.$$

Show that, written out, it has the form $At + Bt^2 = 0$, whence $t = -B/A$. Use this idea to generate solutions to the equation $x^3 + y^3 + z^3 = w^3$ that depend on the parameters u and v, by taking

(a) $(a, b, c, d) = (3, 4, 5, 6)$ and $(p, q, r, s) = (u, -u, v, v)$;

(b) $(a, b, c, d) = (6, 1, 8, 9)$ and $(p, q, r, s) = (u, -u, v, v)$.

29. *High school algebra; moderate difficulty.* S. Brudo has a technique for obtaining additional sets of equal power sums from a given one. To provide a flavor of his approach, begin with the equation $9^3 + 10^3 = 12^3 + 1^3$. Determine p, q, r, s to satisfy the system of equations

$$p + q = 9, \quad r + s = 10, \quad r + q = 12, \quad p - s = 1.$$

Observe that the equation in t,

$$0 = (p + tq)^3 + (r + ts)^3 - (r + tq)^3 - (p - ts)^3$$

has three roots, 0, 1, and a remaining root which is rational. Determine this root and, thence, find another solution of the diophantine equation $x^3 + y^3 = u^3 + v^3$.

30. *Moderate arithmetic.* By factoring the numbers in the following equations as products of smaller integers, verify that $18^3 + 3^6 = 3^8$, $28^2 + 2^9 = 6^4$, $648^2 + 108^3 = 6^8$, and $110592^2 + 4608^3 = 24^8$.

Notes

A good general reference for the older material in this chapter is L. E. Dickson, *History of the Theory of Numbers*, Volume II. Chelsea, 1952. Specific topics can be located as follows:

Pythagorean triples: Chapter IV

Sums of consecutive squares equal to squares: p. 322

Equal sums of two cubes: Chapter XXI, pp. 560–563

Sums of cubes in arithmetic progression: Chapter XXI, pp. 582–585

Consult also Robert D. Carmichael, *The theory of numbers/Diophantine analysis*, Dover, 1959

1. See, for example,

R. Creighton Buck, Sherlock Holmes in Babylon, *American Mathematical Monthly* 87 (1980), 335–345

Matthew Linton, Babylonian triples, *Bulletin of the Institute of Mathematics and its Applications* 24 (1988), 37–41

2. It is readily checked that $(k(m^2 - n^2), 2kmn, k(m^2 + n^2))$ is a pythagorean triple. On the other hand, given a pythagorean triple, we can divide each term by the greatest common divisor k of the three terms to get a triple (x, y, z) for which the greatest common divisor of the terms is 1. x and y cannot have the same parity, so we may supppose that x is odd and y is even; it follows that z also is odd. Rewrite the pythagorean equation as $y^2 = (z - x)(z + x)$. Any common divisor of the even numbers $z - x$ and $z + x$ must divide their sum $2z$ and difference $2x$; it follows that the greatest common divisor of $z - x$ and $z + x$ is 2, and so, $z - x = 2n^2$, $z + x = 2m^2$ for some integers m and n. Hence $x = m^2 - n^2$, $y = 2mn$ and $z = m^2 + n^2$.

In his article in *American Mathematical Monthly* 58 (1951), 30–32, P. J. Schillo addresses the problem of approximating any right triangle up to similarity by one with integer side lengths. In Problem 896, *Mathematics Magazine* 47 (1974), 106; 48 (1975), 119, it is pointed out that if $(a, a+1, c)$ is a pythagorean triple, then so also is $(2c+3a+1, 2c+3a+2, 3c+4a+2)$. A general reference on pythagorean triples is the book, W. Sierpinski, *Pythagorean triples,* Yeshiva, 1962.

3. Problem 1422 in *Journal of Recreational Mathematics* 17 (1984–1985), 218; 18 (1985–1986), 309. Simply take $a = z - y$, $b = x + y - z$, and $c = z - x$ and check that $a^2 + b^2 + c^2 = (x + y - 2z)^2$.

4. Problem E812 in *American Mathematical Monthly* 55 (1948), 248; 56 (1949), 32–33. Given a triple $(k(m^2 - n^2), 2kmn, k(m^2 + n^2))$, with $m > n$, we find that the sum of its members is $2km(m+n)$ which lies between $2km^2$ and $4km^2$. Accordingly, if we pick a perimeter p, we hunt for a factorization $2km(m+n)$ of p by trying m and k for which $2km^2 < p < 4km^2$, and then determining n to suit.

5. Call a pythagorean triple *primitive* if the greatest common divisor of its elements is 1; thus, $(3, 4, 5)$ is primitive; $(6, 8, 10)$ is not. We can identify a structure for the family of primitive pythagorean triples (x, y, z) by noting that $(x/z)^2 + (y/z)^2 = 1$ so that x/z and y/z can be regarded as the cosine and sine of some angle θ. Indeed, select θ so that $\tan(\theta/2) = n/m$, where m and n are integers with greatest common divisor 1. Then

$$\cos \theta = 2 \cos^2 \frac{\theta}{2} - 1 = \frac{2}{1 + \tan^2(\theta/2)} - 1 = \frac{m^2 - n^2}{m^2 + n^2}$$

$$\sin \theta = 2 \sin \frac{\theta}{2} \cos \frac{\theta}{2} = \frac{2 \tan(\theta/2)}{1 + \tan^2(\theta/2)} = \frac{2mn}{m^2 + n^2},$$

so that θ corresponds to the primitive triple $(m^2 - n^2, 2mn, m^2 + n^2)$.

If we think of an angle as representing a rotation of the unit circle about its center, then the superposition of two rotations corresponds to the addition of the corresponding angles. We can "lift" this operation to the corresponding pythagorean triples. Thus, if $\theta \sim (x, y, z)$ and $\psi \sim (u, v, w)$, so that $\cos \theta = x/z$ and $\cos \psi = u/w$, then

$$\cos(\theta + \psi) = \cos \theta \cos \psi - \sin \theta \sin \psi = \frac{xu - yv}{zw}$$

$$\sin(\theta + \psi) = \cos \theta \sin \psi + \cos \psi \sin \theta = \frac{xv + yu}{zw}.$$

$(xu - yv, xv + yu, zw)$ is indeed a pythagorean triple since

$$(zw)^2 = (x^2 + y^2)(u^2 + v^2) = (xu - yv)^2 + (xv + yu)^2.$$

It is natural to define an operation on primitive pythagorean triples (of not necessarily positive integers) that corresponds to addition of angles:

$$(x, y, z) * (u, v, w) = \left(\frac{xu - yv}{d}, \frac{xv + yu}{d}, \frac{zw}{d} \right)$$

where d is the greatest common divisor of its three numerators. For example,

$$(3, 4, 5) * (3, 4, 5) = (-7, 24, 25)$$
$$(3, 4, 5) * (7, 24, 25) = (-3, 4, 5)$$
$$(3, 4, 5) * (-7, 24, 25) = (-117, 44, 125)$$
$$(13, 84, 85) * (3, 4, 5) = (-297, 304, 425)$$
$$(13, 84, 85) * (4, 3, 5) = (-8, 15, 17)$$
$$(84, 13, 85) * (3, 4, 5) = (8, 15, 17)$$
$$(84, 13, 85) * (4, 3, 5) = (297, 304, 325).$$

See, for example, the article by Ernest J. Eckert in *Mathematics Magazine* 57 (1984), 22–27. For another look at the density of rational pythagorean triples, consult the article by L. H. Lange and D. E. Thoro in *American Mathematical Monthly* 71 (1964), 664–665.

6. The nth equation involves $n + 1$ terms on the left and n terms on the right. Let $r = 2n^2 + 2n = 2n(n + 1)$ be the largest term on the left side. We find that the difference of the two sides is

$$\left[(r + 1)^2 + (r + 2)^2 + \cdots + (r + n)^2 \right]$$
$$- \left[(r - 1)^2 + (r - 2)^2 + \cdots + (r - n)^2 \right] - r^2$$
$$= \left[\sum_{i=1}^{n} (r + i)^2 - (r - i)^2 \right] - r^2$$
$$= \left[4r \sum_{i=1}^{n} i \right] - r^2 = r \left[2n(n + 1) - r \right] = 0.$$

See Classroom Note 242 in *Mathematical Gazette* 55 (1971), 54–55 and Problem 550 in *Mathematics Magazine* 37 (1964), 119, 359.

    H. L. Alder in *American Mathematical Monthly* 69 (1962), 282–285; 71 (1964), 749–754 discusses when the sum of n consecutive squares is equal to the sum of $n + k$ consecutive squares. Some possibilities are given by the quartets $(k, n; x, y) = (6, 5; 28, 15), (8, 3; 137, 67), (9, 3; 23, 6), (10, 5; 25, 8),$ $(46, 1; 3854, 539), (96, 1; 679, 15),$ corresponding to the equality of n squares beginning with x^2 and $n + k$ squares beginning with y^2. There are certain values of k for which no solutions exist, such as $2 \leq k \leq 5$ and $k \equiv 3, 4, 5$ (mod 8).

7. These patterns are given in *Mathematical Gazette* 68 (1984), 190 as investigations for students attending a sixth form conference. The first set of equations exemplifies the algebraic equation

$$\sum_{r=n^2}^{n^2+n} r = \sum_{r=n^2+n+1}^{n^2+2n} r$$

both sides of which are equal to $n(n + 1)(2n + 1)/2$, or three times the sum of the first n squares. A "proof without words" can be found in *Mathematics Magazine* 63 (1990), 25.

In Note 2047 in *Mathematical Gazette* 33 (1949), 41–43, D. F. Ferguson shows that, if the sum of $n + r$ consecutive integers is equal to the sum of the next n integers, then $(r, n) = (k^2p, mkp)$ where kp is odd, or $(r, n) = (2k^2p, mkp)$ where mp is odd, and studies other interesting features of this situation. See also Exercise 1 in Chapter 4.

The second set of equations exemplifies

$$\sum_{r=n^2+1}^{(n+1)^2} r = n^3 + (n + 1)^3.$$

8. In general, $n^2 + (n + 1)^2 + [n(n + 1)]^2 = [n(n + 1) + 1]^2$.

See Problem 4511 in *American Mathematical Monthly* 59 (1952), 640; 61 (1954), 130–131. Consider the equation $x^2 + y^2 + (xy)^2 = z^2$, or, equivalently, $(x^2 + 1)(y^2 + 1) = (z^2 + 1)$. If (x, y, z) is a solution, then so also is $(x, (2x^2 + 1)y + 2xz, 2x(x^2 + 1)y + (2x^2 + 1)z)$. We can start with the obvious solutions $(0, k, k)$ and $(k, 0, k)$ for integers k and construct from them the complete infinite set of solutions. A letter in *Mathematical Gazette* 56 (1972), 132 describes how the sum of three consecutive squares can be written in other ways as the sum of three squares.

9. See problem 1692 in *Journal of Recreational Mathematics* 21 (1989), 68; 22 (1990), 74–76. In *American Mathematical Monthly* 5 (1898), 214, it is noted that 39^2 can be written in seven distinct ways as the sum of three squares. Several parametric solutions of $x^2 + y^2 + z^2 = u^2 + v^2 + w^2$ are given in *American Mathematical Monthly* 6 (1899), 17–18.

10. See R. D. Carmichael, *Diophantine equations* (Dover, 1959), 38–43, and *American Mathematical Monthly* 24 (1917), 393.

11. See Dickson's *History*, Volume II, p. 260, 265.

12. In general

$$(2n)^2 + (2n+1)^2 + (2n+2)^2 + (6n^2 + 6n + 2)^2 = (6n^2 + 6n + 3)^2.$$

13. See *Journal of Recreational Mathematics* 22 (1990), 38.

14. See *American Mathematical Monthly* 1 (1894), 256–259. Many such relations can be found using the fact that any number that is odd or divisible by 4 can be expressed as the difference of two squares. If the sum of the first r squares is of either type, then it can be augmented by a further square to give a square result.

15. See problem 6552 in *American Mathematical Monthly* 94 (1987), 694; 97 (1990), 622–624. Laurent Beeckmans in a more recent article (*American Mathematical Monthly* 101 (1994), 437–442) provides some necessary conditions that r must satisfy and gives a complete list of values not exceeding 1000. In *Mathematics Magazine* 48 (1975), 203–207, John Ewell studies the possible values of sums of consecutive numbers and consecutive squares, not necessarily beginning from 1.

16. Problem 533 in *College Mathematics Journal* 25 (1994), 334; 26 (1995), 333. Using the formula for the sum of squares, we find that

$$\left[1^2 + 2^2 + \cdots + n^2\right]\left[(n+1)^2 + (n+2)^2 + \cdots + (2n)^2\right]$$
$$= \frac{n^2(2n+1)^2}{36}(n+1)(7n+1)$$

which is square exactly when $(n+1)(7n+1) = m^2$ for some integer m. This is equivalent to $(7n+4)^2 - 7m^2 = 9$. We need to solve the diophantine equation $r^2 - 7m^2 = 9$ for $r \equiv 4 \pmod 7$. Given any solution (r_0, m_0), we can generate infinitely many others using the recursion

$$r_{k+1} = 8r_k + 21m_k$$
$$m_{k+1} = 3r_k + 8m_k.$$

By inspection, we obtain the solutions $(r, m) = (4, 1)$ and $(11, 4)$. From these, with $r = 7n + 4$, we get as possibilities for (r, m, n):

$(4, 1, 0)$ extraneous, $(53, 20, 7), (844, 319, 120), (13451, 5084, 1921),$

and

$$(11, 4, 1), (172, 65, 24), (2741, 1036, 391).$$

By the way, note that $(1^2 + 2^2)(3^2 + 4^2) = 5^3$. Are there any other values of n for which $[1^2 + \cdots + n^2][(n+1)^2 + \cdots + (2n)^2]$ is a cube?

17. See problem E3064 in *American Mathematical Monthly* 91 (1984), 649; 94 (1987), 190–192.

18. The sum of three consecutive cubes is always divisible by 3, so the condition of three consecutive cubes summing to a square leads to the diophantine equation

$$(n-1)^3 + n^3 + (n+1)^3 = (3m)^2$$

or, more succinctly,

$$n(n^2 + 2) = 3m^2.$$

The two examples given are exhaustive; there are no more. For a detailed discussion, see the comments on Problem 67.B in *Mathematical Gazette* 67 (1983), 228–230.

Similarly, the diophantine equation for five consecutive cubes summing to a square is

$$n(n^2 + 6) = 5m^2.$$

Both of these equations are not easy to solve, and indeed will have only finitely many solutions. For example, suppose we are searching for an odd value of n for which $n(n^2 + 2) = 3m^2$. Since n and $n^2 + 2$ are coprime and since $n^2 + 2$ cannot be a square (no two squares can differ by 2), we must have $n = r^2$ and $n^2 + 2 = 3s^2$ for some integers r and s. Thus, one needs to search among the solutions to the pellian equation $n^2 - 3s^2 = -2$ for all of those for which n is a square, not an easy task.

19. This failure of a cube to be expressible as the sum of two nonzero cubes gives rise to a little whimsy. In *Mathematics Magazine* 63 (1990), 55, we have the "counterexample" $(x, y, z) = (3 + \sqrt{93}, \ 3 - \sqrt{93}, \ 12)$. In *Mathematical Gazette*, M. Rumney recounts how he thought he had shown that $24^3 = 31^3 + 11^3$. He actually had $13 \times 31^2 + 11 \times 11^2 = 24^3$ and asks whether there are further solutions of $ux^2 + vy^2 = z^3$, where u and v have, respectively, the same digits as x and y, but in reverse order.

20. Consult Ramanujan's *Lost Notebook*, p. 341. This, of course, is not the only word of Ramanujan on this sort of thing. See the articles by T. S. Nanjundiah in *American Mathematical Monthly* 100 (1993), 485–486 and by B. C. Berndt and S. Bhargava in *American Mathematical Monthly* 100 (1993), 644–656. For example, if $\alpha^2 + \alpha\beta + \beta^2 = 3\lambda\gamma^2$ (satisfied, for example, by $(\alpha, \beta, \gamma, \lambda) = (3, 0, 1, 3)$), then

$$(\alpha + \lambda^2\gamma)^3 + (\lambda\beta + \gamma)^3 = (\lambda\alpha + \gamma)^3 + (\beta + \lambda^2\gamma)^3.$$

Also,

$$(8s^2 + 40st - 24t^2)^4 + (6s^2 - 44st - 18t^2)^4 + (14s^2 - 4st - 42t^2)^4$$
$$+ (9s^2 + 27t^2)^4 + (4s^2 + 12t^2)^4 = (15s^2 + 45t^2)^4$$

and

$$(4m^2 - 12n^2)^4 + (3m^2 + 9n^2)^4 + (2m^2 - 12mn - 6n^2)^4 +$$
$$(4m^2 + 12n^2)^4 + (2m^2 + 12mn - 6n^2)^4 = (5m^2 + 15n^2)^4.$$

If $ad = bc$, then, for $n = 2$ or 4,

$$(a+b+c)^n + (b+c+d)^n + (a-d)^n = (c+d+a)^n + (d+a+b)^n + (b-c)^n.$$

21. See *American Mathematical Monthly* 100 (1993), 331–340, as well as Problem 1298 in the *Journal of Recreational Mathematics* 17 (1984–1985), 148–151 for other examples.

22. There is, however, a more complicated generalization provided by A. H. Stone and J. E. Simpson in two articles in the *American Mathematical Monthly*, respectively 92 (1985), 328–331 and 93 (1986), 701–708.

23. See Problem 86 in *American Mathematical Monthly* 9 (1902), 79. The method suggested in the exercise for this note for generating other two-parameter families is given in the little book, *Algebra can be fun*, by Ya. I. Perelman (Mir, Moscow, 1979), on pages 139–143. The last solution (VIII) is due to Ramanujan (email message from Richard Askey, July, 1996); it can be conveniently checked by factoring $w^3 - x^3$ and $y^3 + z^3$. In June, 1995, Vivakanand Kadarnauth found the related examples:

$$4^3 + 5^3 + 3^3 = 6^3 \qquad\qquad 4^3 + 17^3 + 22^3 = 25^3$$
$$16^3 + 23^3 + 41^3 = 44^3 \qquad\qquad 16^3 + 47^3 + 108^3 = 111^3$$
$$64^3 + 107^3 + 405^3 = 408^3 \qquad\qquad 64^3 + 155^3 + 664^3 = 667^3$$

(written communication and an advertisement in the Toronto *Globe and Mail,* January 6, 1996). From the data, we see that there are apparently two solutions with $x = 4^n$, say, (x, y_1, z_1, w_1) and (x, y_2, z_2, w_2) with $y_1 > y_2$. We conjecture that $y_1 + y_2 = 4^{n+1} + 6$ and $y_1 - y_2 = 6 \cdot 2^n$, as well as $w_i = z_i + 3$. Following up on this yields the quartets

$$(x, y, z, w) =$$
$$\left(4^n, \ 2 \cdot 4^n + 3 \cdot 2^n + 3, \ 2^n(4^n + 2^{n+1} + 3), \ 2^n(4^n + 2^{n+1} + 3) + 3\right)$$

and

$$(x, y, z, w) =$$
$$\left(4^n, \ 2 \cdot 4^n - 3 \cdot 2^n + 3, \ 2^n(4^n - 2^{n+1} + 3) - 3, \ 2^n(4^n - 2^{n+1} + 3)\right).$$

24. In *Mathematical Gazette* 21 (1937), 33–35, A. Russell and C.E. Gwyther provide a list of representations of each cube up to 200^3 as the sum of three cubes. The solution to Problem E1249 in *American Mathematical Monthly* 64 (1957), 43, 507–508 lists all cubes up to 1000^3 that can be written as the sum of three cubes in at least five ways.

25. See Dickson's *History*, Volume II, pages 582–585.

26. See T. J. Lander, T. R. Parkin, and J. L. Selfridge, *Mathematics of Computation* 21 (1967), 446–459.

27. T. J. Lander and T. R. Parkin, *Mathematics of Computation* 21 (1967), 101–103.

T. J. Lander and T. R. Parkin, *Bulletin of the American Mathematical Society* (1) 72 (1966), 1079

28. See Noam D. Elkies, *Mathematics of Computation* 51 (1988), 825–835. Elkies' result plays a role in one of the solutions to the problem of finding triangles whose side lengths and areas are all integers (Problem 6628 in *American Mathematical Monthly* 97 (1990), 350; 98 (1991), 772–774). If (x, y, z, w) is a quartet of positive integers whose greatest common divisor is 1 and for which $x^4 + y^4 + z^4 = w^4$, then the triangle whose sides are $y^4 + z^4$, $z^4 + x^4$, $x^4 + y^4$ has area $(xyzw)^2$. A brief article on the work of Frye and Elkies by Barry Cipra appears in *Science* 239 (1988), 464. The results are also mentioned in *American Mathematical Monthly* 96 (1989), 906.

29. See the articles by S. Brudno in *Mathematics of Computation* 23 (1969), 877–880; 24 (1970), 453–454, in which the author discusses how one instance of equal power sums can be used to generate others. The example in the exercise for this note leads to $t = 5/3$, from which the equation

$3^3 + 36^3 + 37^3 = 46^3$ can be derived. A corrigendum appears in *Mathematics of Computation* 25 (1971), 409.

In *American Mathematical Monthly* 75 (1968), 1061–1073, L. J. Lander studies equal sums of like powers, in particular, sums of two fourth powers and of three and four fifth powers.

30. See the expository articles by Cox and by Goevêa in *American Mathematical Monthly* 101 (1994), 3–14, 203–222.

31. For example, we can check the last equation in the list of five by

$$110592^2 + 4608^3 = (2^{12} \times 3^3)^2 + (2^9 \times 3^2)^3 = (2^{24} \times 3^6) + (2^{27} \times 3^6)$$

$$= 2^{24} \times 3^6 \times (1 + 2^3) = 2^{24} \times 3^8 = (2^3 \times 3)^8 = 24^8.$$

See the article by Kenneth M. Hoffman in *Math Horizons* (September, 1994), 29; this is a magazine for undergraduates published by the Mathematical Association of America. In *American Mathematical Monthly* 95 (1988), 544–547, David W. Boyd shows that $x^2 + y^m = z^{2m}$ has infinitely many primitive solutions when m and n are relatively prime.

Additional Reading

R. D. Carmichael, On the impossiblity of certain diophantine equations and systems of equations. *American Mathematical Monthly* 20 (1913), 213–221.

Some results: There do not exist squares whose sum and difference are both squares. The area of a right triangle is never a square nor twice a square. No square is the difference of two fourth powers.

R. D. Carmichael, On certain diophantine equations having multiple parameter solutions. *American Mathematical Monthly* 20 (1913), 304–307.

Andrew Bremner and Richard K. Guy, A dozen difficult diophantine dilemmas. *American Mathematical Monthly* 95 (1988), 31–36.

A discussion of problems, including the determination of numbers expressible as the sums of the same powers of two distinct pairs.

Additional Exercises

1. Check the algebraic identities given in this chapter.

   2. Verify that, for each integer n,

$$\left(2^{4n} + 2^{2n+1}, 2^{4n} - 2^{4n-2} - 2^{2n} - 1, 2^{4n} + 2^{4n-2} + 2^{2n} + 1\right)$$

is a pythagorean triple. This was found in *American Mathematical Monthly* 57 (1950), 331–332, where a "colossal" primitive triple is given.

3. Consider the following table with respect to right triangles:

Pythagorean triple	semiperimeter	area
$(3, 4, 5)$	$1 + 2 + 3$	6×1^2
$(5, 12, 13)$	$1 + 2 + 3 + 4 + 5$	$6 \times (1^2 + 2^2)$
$(7, 24, 25)$	$1 + \cdots + 7$	$6 \times (1^2 + 2^2 + 3^2)$

Conjecture and establish a generalization.

4. Determine the general solution of $x^2 + y^2 = z^4$ in positive integers x, y, z whose greatest common divisor is 1.

5. If the smallest two numbers of a pythagorean triple add up to a square, show that the sum of their cubes is equal to the sum of two squares.

6. Note the "almost pythagorean" equations

$$10^2 + 15^2 = 18^2 + 1 \quad 20^2 + 25^2 = 32^2 + 1 \quad 25^2 + 35^2 = 43^2 + 1.$$

(a) Verify that $(x, y, z) = (3t + 2, 4t + 1, 5t + 2)$ and $(x, y, z) = (3t + 1, 4t + 3, 5t + 3)$ satisfy $x^2 + y^2 = z^2 + 1$.

(b) Indicate ways of finding infinitely many solutions in integers of each of the equations $x^2 + y^2 = z^2 + 1$ and $x^2 + y^2 = z^2 - 1$.

7. (a) Find all triples of numbers (x, y, z) with greatest common divisor 1 for which

$$\frac{1}{x^2} + \frac{1}{y^2} = \frac{1}{z^2}.$$

(b) Determine three squares in harmonic progressions, i.e., numbers whose reciprocals are in arithmetic progression.

8. Determine numbers x, y, z for which

$$x^2 + y^2 + 3z^2 = (x + y + z)^2.$$

9. Find triples (x, y, z) of numbers the sum of whose cubes is divisible by their product. Are there infinitely many such numbers?

10. If we drop the condition that the greatest common divisor of x, y, z is 1, then it becomes much easier to find solutions for the diophantine equation $x^p + y^q = r^q$ and its analogues with more variables. Determine solutions to
(a) the "near" Fermat equation $x^n + y^n = z^{n+1}$;
(b) $x^2 + y^3 = z^4$;
(c) $x^3 + y^4 = z^5$;
(d) $x^3 + y^5 = z^7$;
(e) $x^4 + y^7 + z^9 = w^{11}$.

11. Determine all triples (x, y, z) of natural numbers for which

$$3^x + 4^y = 5^z.$$

12. (a) Prove that, if p is prime, then $2^p + 3^p$ is not a power of an integer.

(b) Find all pairs (m, n) of natural numbers for which $2^m + 3^n$ is a perfect square.

13. Find all triples (x, y, z) of nonnegative integers for which $4^x + 4^y + 4^z$ is the square of an integer.

14. (a) Show that, when $n > 2$, the Fermat equation $x^n + y^n = z^n$ has no solutions in positive integers x, y, z for which

$$z < \frac{2^{1/n}}{2^{1/n} - 1}.$$

(b) In particular, when $n = 10$, show that there are no solutions with $z \leq 13$.

15. Find all solutions in positive integers a, b, c, d for the exponential equation

$$1 + 3^a = 5^b + 3^c.$$

Solutions

3. If the longest sides of the triangle differ in length by 1, then the corresponding pythagorean triple is $(2k + 1, 2k^2 + 2k, 2k^2 + 2k + 1)$. The semiperimeter is $2k^2 + 3k + 1 = \frac{1}{2}(2k + 1)(2k + 2) = 1 + 2 + \cdots + (2k + 1)$ and the area is $k(k + 1)(2k + 1) = 6(1^2 + 2^2 + \cdots + k^2)$, which, as we have seen, is square when $k = 1$ and $k = 24$. According to Dickson's *History* (Volume II, pp. 181–183), it is not possible for the area of a right triangle with integer sides to be of the form rn^2 for $1 \leq r \leq 5$. This problem is from *Mathematics Teacher* 87 (1994), 662. Problem 1088 in *Mathematics Magazine* 54 (1987), 36 asks, for given positive integer n, how many pythagorean triples correspond to triangles whose area is n times the perimeter.

4. See *American Mathematical Monthly* 20 (1913), 313; 21 (1914), 199–200. We have, in particular, $7^2 + 24^2 = 5^4$, $119^2 + 120^2 = 13^4$, and $161^2 + 240^2 = 17^4$.

5. Let $c^2 = a^2 + b^2$. Then

$$a^3 + b^3 = (a + b)(a^2 - ab + b^2)$$
$$= (a + b)\left[\left[\tfrac{1}{2}(c + b - a)\right]^2 + \left[\tfrac{1}{2}(c - b + a)\right]^2\right]$$

from which the result follows. In the published solution of this problem (4205 in *American Mathematical Monthly* 53 (1946), 278; 54 (1947), 550), it is noted that the hypotenuse itself can be a square. The smallest example of such a pythagorean triple is $(4565486027761, 1061652293520, 4687298610289)$. The sum of the smallest two numbers is 2372159^2 and the largest number is 2165017^2.

6. See *Mathematics Magazine* 60 (1987), 234–236, 244 (Q724).

7. (a) For any such triple, (zy, zx, xy) is a pythagorean triple. Using this fact, we can determine that

$$(x, y, z) = \big(2mn(m^2 + n^2), (m^2 - n^2)(m^2 + n^2), 2mn(m^2 - n^2)\big)$$

for some integers m and n. (*American Mathematical Monthly* 57 (1950), 334; 58 (1951), 41)

(b) See *American Mathematical Monthly* 7 (1900), 82–83; 9 (1902), 79–80. Three such squares are 25, 49 and 1225.

8. The equation can be rewritten $2z^2 = (z + x)(z + y)$. If the greatest common divisor of x, y, z is equal to 1, we can argue that z must have the same parity as at least one of the other variables, say x. Then, for some integers m and n, we must have $z + x = 2m^2$ and $z + y = n^2$ from which a parametric solution is found. The general solution is

$$(x, y, z) = \big(km(2m - n), kn(n - m), kmn\big)$$

where $m < n < 2m$. (Problem 941 in *American Mathematical Monthly* 57 (1950), 686; 58 (1951), 491–492)

9. We wish to find positive numbers p, q, r, n for which

$$p^3 + q^3 + r^3 = npqr.$$

Some possibilities are

$$(p, q, r; n) = (1, 1, 2; 5), \ (1, 2, 3; 6), \ (2, 3, 7; 9), \ (1, 5, 9; 19),$$

$$(1, 2, 9; 41), \ (1, 5, 14; 41), \ (1, 3, 14; 66).$$

Ben Dushnik in *American Mathematical Monthly* 53 (1946), 451–452 provides ways of generating new solutions from existing ones. In particular, if $(p, q, r) = (1, y, z)$ works for some n, then so does $(p, q, r) = (y^2 + z^2 - yz, y, z)$.

The cubic equation arises in the problem of determining integers x, y, z for which

$$\frac{y}{z} + \frac{z}{x} + \frac{x}{y}$$

is an integer (Problem E682 in *American Mathematical Monthly* 52 (1945), 395; 53 (1946), 223–224). Simply try $x = pr^2$, $y = p^2q$, $z = q^2r$.

10. (a) $(x, y, z) = (2, 2, 2)$.

(b) Use the factorization $y^3 = (z^2 - x)(z^2 + x)$ to generate some values of z and x for given y. To begin with, try $z^2 - x = y$ and $z^2 + x = y^2$, which leads to $2z^2 = y(y+1)$; see Chapter 4, applications of Pell's equation. In *American Mathematical Monthly* 95 (1988), 544–547, D. W. Boyd shows that $x^2 + y^m = z^{2n}$ is always solvable when m and n are relatively prime, and gives a method for solving them.

(c) See Quickie Q403 in *Mathematical Magazine* 40 (1967), 30, 54.

(d) This problem appeared in *Mathematics and Informatics Quarterly* 4 (1994), 211–212; it was on Round 12 of the International Mathematical Talent Search for high school students. Some of the solutions given for (x, y, z) are:

(i) $(5 \times 3^{28}, 3^{17}, 2 \times 3^{12})$ (inspired by $5^3 + 3^1 = 2^7$);

(ii) $(a^{24}b^{21}c^{25}, a^{14}b^{13}c^{15}, a^{10}b^9c^{11})$ where $a^2 + b^2 = c^2$;

(iii) $(ac^{30}, bc^{18}, c^{13})$ where $c = a^3 + b^5$;

(iv) $(2^{30}m^{35}, 2^{18}m^{21}, 2^{13}m^{15})$;

(v) $(m(m^3 + n^5)^{30}, n(m^3 + n^5)^{18}, (m^3 + n^5)^{13})$;

(vi) $(2^{35k+30}, 2^{21k+18}, 2^{15k+13})$;

(vii) $(2 \times 3^{35k+25}, 3^{21k+15}, 3^{15k+11})$;

(viii) $(2^{35k+14} \times 3^{35j+25}, 2^{21k+9} \times 3^{21j+15}, 2^{15k+6} \times 3^{15j+11})$.

Note that, if (x, y, z) is a solution, then so also is $(m^{35}x, m^{21}y, m^{15}z)$.

(e) This is a "Quickie", Q795 in *Mathematics Magazine* 65 (1992), 266, 273. Try $(x, y, z, w) = (3^a, 3^b, 3^c, 3^d)$ where $4a = 7b = 9c$ and $d = (4a + 1)/11$.

11. This is Problem 1369 in *Mathematics Magazine* 64 (1991), 131; 65 (1992), 131–132.

Since $1 \equiv 5^z \equiv 3^x \equiv (-1)^x \pmod 4$ and $1 \equiv 4^y \equiv 5^z \equiv (-1)^z \pmod 3$, x and z must be even. Let $x = 2u$ and $z = 2w$. Then

$$4^y = (5^w - 3^u)(5^w + 3^u)$$

so that $5^w - 3^u = 2^r$ and $5^w + 3^u = 2^s$ for positive integers r, s with $1 \le r < s$ and $r + s = 2y$. Hence

$$5^w = 2^{r-1}(1 + 2^{s-r}) \quad \text{and} \quad 3^u = 2^{r-1}(2^{s-r} - 1).$$

It follows that $r = 1$ and so $5^w = 3^u + 2$, whence

$$2^{s-1} = 5^w - 1 = 4(5^{w-1} + 5^{w-2} + \cdots + 5 + 1)$$

or

$$2^{s-3} = 5^{w-1} + 5^{w-2} + \cdots + 5 + 1.$$

This is satisfied if $s = 3$ and $w = 1$, so that $(x, y, z) = (2, 2, 2)$. If $s > 3$, then w must be even and we obtain

$$2^{s-3} = (1 + 5) + (5^2 + 5^3) + \cdots + (5^{w-2} + 5^{w-1})$$
$$= 6(1 + 5^2 + 5^4 + \cdots + 5^{w-2}),$$

i.e., a power of 2 is a multiple of 6, an impossibility. Hence $(x, y, z) = (2, 2, 2)$ is the only possibility.

This problem is considered in W. Sierpinski, *Elementary theory of numbers*, 1988 (page 40).

12. Problem E2396 in *American Mathematical Monthly* 80 (1973), 76; 81 (1974), 172–173.

(a) The result is clear if $p = 2$ or $p = 5$. Otherwise, p is odd and $2^p + 3^p = 5(2^{p-1} - 2^{p-2} \times 3 + \cdots + 3^{p-1})$. The second factor has oddly many terms and, being congruent to $p2^{p-1}$ modulo 5, is not divisible by 5. The result follows.

(b) Let $t^2 = 2^m + 3^n$, if possible. Then t is odd and we see, from reducing modulo 3 and 4, that m and n are both even. Let $n = 2r$. Then $2^m = (t + 3^r)(t - 3^r)$. The greatest common divisor of the two factors must be 2, so that $t - 3^r = 2$ and $t + 3^r = 2^{m-1}$. Hence $3^r + 1 = 2^{m-2}$. Now, $m = 2$ is not possible and $m = 4$ leads to $5^2 = 2^4 + 3^2$. If $m \geq 6$, then 8 must divide $3^r + 1$, an impossibility. Hence $(m, n) = (4, 2)$ is the sole acceptable pair.

13. Problem E2075 in *American Mathematical Monthly* 75 (1968), 403; 76 (1969), 308–309. Suppose that $4^x + 4^y + 4^z$, with $x \leq y \leq z$, is a square. Dividing the expression by the square 4^x yields the relation

$$1 + 4^u + 4^v = (1 + 2^m t)^2 = 1 + 2^{m+1} t + 4^m t^2$$

for nonnegative integers $u = y - x$, $v = z - x$, m and the odd positive integer t.

It is clear that $m \neq 0$. Suppose $m = 1$; then we are led to

$$4^{u-1}(1 + 4^{v-u}) = t(t + 1)$$

which is solvable with $v = 2u - 1$ and $t = 4^{u-1}$.

If $m \geq 2$, then we are led to

$$4^u(1 + 4^{v-u}) = 2^{m+1} t(1 + 2^{m-1} t).$$

We must have that $2u = m + 1$. Setting $w = v - u$, we find that

$$t - 1 = 4^w - 2^{m-1} t^2 = 2^{2w} - 2^{2u-2} t^2$$
$$= 4^{u-1}(2^{w-u+1} + t)(2^{w-u+1} - t).$$

If $t > 1$, we get a contradiction since the right side exceeds t and the left side does not. Hence $t = 1$ and $w - u + 1 = 0$. Therefore, $v = 2u - 1$, and so $z = 2y - x - 1$. Indeed, we find that, for all nonnegative integers x and y, $4^x + 4^y + 4^{2y-x-1} = (2^x + 2^{2y-x-1})^2$.

14. See problem E1289 in *American Mathematical Monthly* 64 (1957), 671; 65 (1958), 369.

(a) Suppose that $x^n + y^n = z^n$. Since $0 < x, y < z$, we must have that $x, y \leq z - 1$, whence $z^n \leq 2(z-1)^n$, or $z \leq 2^{1/n}(z-1)$. This occurs if and only if $z \geq 2^{1/n}/(2^{1/n} - 1)$.

(b) Observe that, from the Binomial Theorem,

$$\left(1 + \frac{1}{12}\right)^{10} > 1 + \frac{5}{6} + \frac{45}{144} > 2,$$

whence

$$1 < 2^{1/10} < 13/12 \implies 1 > 2^{-1/10} > 12/13$$

$$\implies 0 < 1 - 2^{-1/10} < 1/13$$

$$\implies \frac{2^{1/10}}{2^{1/10} - 1} = \frac{1}{1 - 2^{-1/10}} > 13.$$

From this, we deduce that any solution of $x^{10} + y^{10} = z^{10}$ must have $z \geq 14$.

15. (a) Problem E2929, *American Mathematical Monthly* 89 (1982), 131; 91 (1984), 141. An obvious solution is given by $a = c$, $b = 0$. There are no solutions with $b = 1$ and the possibility $b = 2$ leads to the solution $(a, b, c) = (3, 2, 1)$. Suppose that $b \geq 3$. We have that $a > c \geq 1$. Then $1 + 3^a \equiv 3^c \pmod 5$, so that $(a, c) \equiv (0, 3), (1, 2), (3, 1) \pmod 4$. But $3^a \equiv 3^c$ (mod 4), so that a and c must have the same parity. Hence $(a, c) \equiv (3, 1)$ (mod 4), so that $3^a \equiv 11$ and $3^c \equiv 3 \pmod{16}$.

It follows that $5^b \equiv 1 + 3^a - 3^c \equiv 9 \pmod{16}$, whence $b \equiv 2$ (mod 4). Now, modulo 13, $3^3 \equiv 5^4 \equiv 1$, and we find that $2 + 3^a \equiv 3^c$ whence $(a, c) \equiv (0, 1) \pmod 3$. It follows that $(a, c) \equiv (3, 1) \pmod 6$, so that $3^a \equiv -1$ and $3^c \equiv 3 \pmod 7$. Thus, $5^b \equiv 4 \pmod 7$, and so $b \equiv 2 \pmod 6$.

If $c > 1$, $3^a \equiv 3^c \equiv 0 \pmod 9$, whence $1 \equiv 5^b \equiv 7 \pmod 9$, a contradiction. If $c = 1$, then $a \geq 4$, so that $5^b \equiv -2 \pmod{81}$. Hence $b \equiv 20 \pmod{54}$. It follows that $3^a \equiv 5^{20} + 2 \equiv 37 \pmod{109}$, which is impossible. ($3^{27} \equiv 1 \pmod{109}$ and 37 is not a quadratic residue, modulo 109.)

Further problems of this type are Problem 6411 in *American Mathematical Monthly* 89 (1982), 788; 92 (1985), 63 and Problem E3019 in *American Mathematical Monthly* 90 (1983), 644.

CHAPTER **3**

Sequences

Pythagorean triples again

In Chapter 2, it was noted that each odd number was the smallest element of a pythagorean triple whose largest two numbers were consecutive. We can use this fact to build a chain of triples for which the largest number of each is the smallest number of the rest. For example, beginning with $(3, 4, 5)$, we obtain the chain

$$(3, 4, 5) \rightarrow (5, 12, 13) \rightarrow (13, 84, 85) \rightarrow (85, 3612, 3613)$$

$$\rightarrow (3613, 6526884, 6526885) \rightarrow \cdots$$

Let us focus on the even numbers:

$$4 = 2 \times 2$$

$$12 = 3 \times 4 = 2 \times 2 \times 3$$

$$84 = 7 \times 12 = 2 \times 2 \times 3 \times 7$$

$$3612 = 43 \times 84 = 2 \times 2 \times 3 \times 7 \times 43$$

$$6526884 = 1807 \times 3612 = 2 \times 2 \times 3 \times 7 \times 43 \times 1807$$

$$\cdots$$

Each one is a multiple of the previous one, the consecutive multipliers being the numbers

$$2, \quad 3 = 2 + 1, \quad 7 = 2 \times 3 + 1, \quad 43 = 2 \times 3 \times 7 + 1,$$

$$1807 = 2 \times 3 \times 7 \times 43 + 1, \ldots.$$

Each term in the sequence is one more than the product of all its predecessors. There is another way in which we can obtain the same sequence. Start with $1/2$ and continually add the largest integer reciprocal which will keep the sum of the reciprocals obtained so far less than 1:

$$\frac{1}{2} + \frac{1}{2} = 1 \qquad \frac{1}{2} + \frac{1}{3} < 1$$

$$\frac{1}{2} + \frac{1}{3} + \frac{1}{6} = 1 \qquad \frac{1}{2} + \frac{1}{3} + \frac{1}{7} < 1$$

$$\frac{1}{2} + \frac{1}{3} + \frac{1}{6} + \frac{1}{42} = 1 \qquad \frac{1}{2} + \frac{1}{3} + \frac{1}{7} + \frac{1}{43} < 1$$

$$\cdots$$

This sequence of multipliers arises from the pythagorean triples in another way:

$$5 - 3 = 2$$

$$13 - 5 = 12 - 4 = 8 = 2 \times 2^2$$

$$85 - 13 = 84 - 12 = 72 = 8 \times 3^2 = 2 \times 2^2 \times 3^2$$

$$3613 - 85 = 3612 - 84 = 3528 = 72 \times 7^2 = 2 \times 2^2 \times 3^2 \times 7^2$$

$$6526885 - 3613 = 6526884 - 3612 = 3528 \times 43^2 = 2 \times 2^2 \times 3^2 \times 7^2 \times 43^2.$$

The formation of the sequence $\{2, 3, 7, 43, \ldots\}$ is reminiscent of the proof of Proposition 20 in Book IX of Euclid's *Elements* that there is an arbitrarily large number of primes. Suppose that we have already discovered finitely many primes $p, q, \ldots, v$. Then we can always find one more by taking the number $pq \cdots v + 1$, which is the product of all the primes found so far increased by 1, and extracting a prime factor. This prime factor will necessarily be different from what we have so far.

Applying this to 2 as our initial prime, we find the sequence

$$2,$$

$$3 = 2 + 1,$$

$$7 = 2 \times 3 + 1 \quad \text{a prime,}$$

$$43 = 2 \times 3 \times 7 + 1 \quad \text{a prime,}$$

13 the smallest prime factor of $13 \times 139 = 1807 = 2 \times 3 \times 7 \times 43 + 1,$

and so on. If we choose the smallest prime factor at each stage, will we pick up each prime? **(1)**

Returning to the pythagorean triples, we can start with another odd number, 7, and build the chain

$$(7, 24, 25) \rightarrow (25, 312, 313) \rightarrow (313, 48984, 48985) \rightarrow \cdots$$

and note that

$$24 = 2 \times 12; \quad 312 = 2 \times 12 \times 13; \quad 48984 = 2 \times 12 \times 13 \times 157; \ \ldots$$

and

$$313 - 25 = 312 - 24 = 288 = 2 \times 12^2;$$

$$48985 - 313 = 48984 - 312 = 48672 = 2 \times 12^2 \times 13^2; \ \ldots.$$

These relations involve the sequence $\{12, 13 = 12+1, 157 = 12 \times 13 + 1, \ldots\}$.

(2)

We turn from almost equality of the largest two numbers of a pythagorean triple to how close the smallest two numbers can be. As we saw in Chapter 2, there are infinitely many triples whose smallest members differ only by 1: $(3, 4, 5), (20, 21, 29), (119, 120, 169)$. Here is a another way to find them all.

Form the infinite sequence $1, 6, 35, 204, 1189, 6930, 40391, \ldots$, in which each term after the first two is 6 times its immediate predecessor minus the previous term. Thus, we find that

$$35 = 6 \times 6 - 1; \quad 204 = 6 \times 35 - 6;$$

$$1189 = 6 \times 204 - 35; \quad 6930 = 6 \times 1189 - 204; \ldots.$$

Take any two consecutive terms in this sequence. Their difference will be the largest number in the pythagorean triple. Their sum is an odd number, which can be written as the sum of two consecutive integers; these consecutive integers will be the smallest two numbers of the triple. For example,

$$1189 - 204 = 985; \quad 1189 + 204 = 1393 = 696 + 697.$$

A pythagorean triple is $(696, 697, 985)$. Let us check this by using the difference of squares factorization:

$$985^2 - 697^2 = (985 + 697)(985 - 697) = 1682 \times 288$$

$$= 2 \times 841 \times 2 \times 144$$

$$= 2^2 \times 29^2 \times 12^2 = (2 \times 29 \times 12)^2 = 696^2.$$

The reader may wish to produce a few more pythagorean triples by this method.

However, our sequence has by no means exhausted its powers to fascinate. We can produce two other sequences from the factors of its terms:

The sequence	Sequence A	Sequence B
1	1	1
6	2	3
35	5	7
204	12	17
1189	29	41
6930	70	99
40391	169	239

I do not want to spoil the reader's fun by revealing too much, but this situation is replete with interesting relations. Here are a few questions to consider:

1. Each term of Sequence A can be found in a straightforward way from its two immediate predecessors. Do you see how? The same pattern works for Sequence B.
2. Each term of Sequence A can be found in a simple way from the previous terms of Sequences A and B. What is the pattern? There is a slightly more complicated analogous pattern for sequence B.
3. Use only Sequences A and B to find the pythagorean triples whose smallest terms differ by 1.
4. Add the squares of consecutive terms of Sequence A; what do you observe? Do the same for Sequence B.
5. For each of the three sequences, determine the difference between the square of any entry and the product of its two immediate neighbors in the sequence.

There is much more.

The foregoing process can be generalized to yield other pythagorean triples whose smallest numbers have a specified difference. For example, the triples $(5, 12, 13)$ and $(8, 15, 17)$ each have their smallest entries differing by 7. What other triples can we find to give the same difference? If we allow negative integers, an example is $(-4, 3, 5)$. Here is the generating sequence, which we will extend in both directions:

$$\ldots, -117, -20, -3, 2, 15, 88, \ldots.$$

Each term in the sequence is 6 times its immediate predecessor minus the term before, so that, for example, $15 = 6 \times 2 - (-3)$. (If we wish, we can use the same law of formation working from right to left rather than from left to right.)

As before, the difference of two consecutive terms is the largest term of a pythagorean triple; the sum of two consecutive terms can be written as the sum of two numbers which differ by 7 and which constitute the smallest terms of the triple. The chunk of the sequence displayed above yields the triples $(-72, -65, 97)$, $(-15, -8, 17)$, $(-4, 3, 5)$, $(5, 12, 13)$, $(48, 55, 73)$. The reader may wish to extend the sequence and determine some more triples.

In the most general situation, let us look at the pythagorean triple $(m^2 - n^2, 2mn, m^2 + n^2)$ and ask for other triples whose smallest terms have the same difference $2mn - (m^2 - n^2)$. In this case, the generating sequence will contain the following four consecutive terms

$$mn - n^2 = (m - n)n$$

$$m^2 + mn = (m + n)m$$

$$6m^2 + 5mn + n^2 = (3m + n)(2m + n)$$

$$35m^2 + 29mn + 6n^2 = (7m + 3n)(5m + 2n).$$

The second and third terms of the sequence correspond to the triple $(4m^2 + 2mn, 3m^2 + 4mn + n^2, 5m^2 + 4mn + n^2)$. The interested reader will be left with the task of exploring this further.

We can apply this technique to generate solutions to other quadratic equations besides the pythagorean equation $x^2 + y^2 = z^2$. For example, let us solve $x^2 + 2y^2 = z^2$ where we will insist that x and y differ in absolute value by 1. In this case, the generating sequence is

$$1, 4, 15, 56, 209, \ldots.$$

Note that each term of this sequence is 4 times its predecessor minus the term before. Now observe:

$4 - 1 = 3;$	$4 + 1 = 5 = 1 + 2 \times 2;$	$3^2 = 1^2 + 2 \times 2^2$
$15 - 4 = 11;$	$15 + 4 = 19 = 7 + 2 \times 6;$	$11^2 = 7^2 + 2 \times 6^2$
$56 - 15 = 41;$	$56 + 15 = 71 = 23 + 2 \times 24;$	$41^2 = 23^2 + 2 \times 24^2$
$209 - 56 = 153;$	$209 + 56 = 265 = 89 + 2 \times 88;$	$153^2 = 89^2 + 2 \times 88^2$

See Note **(3)** at the end of this chapter.

Second-order recursions

In the last section, we met sequences of integers in which each entry past the second depended in a simple way on the two preceding entries. In talking about sequences of numbers, it is convenient to use a subscripted notation: x_1 refers to the first term in the sequence; x_2 refers to the second term in the sequence;

and where n is any positive integer, we let x_n refer to the nth term of the sequence. Thus, the sequence can be written

$$\{x_1, x_2, x_3, x_4, \ldots, x_n, \ldots\}.$$

(Sometimes, we also use zero or negative integers for subscripts. Thus, a sequence may have a zeroth term x_0 or even an earlier term like x_{-3}.)

A *second-order linear recursion* is a sequence in which the first two terms can be anything you want. After that, there are multipliers a and b such that the succeeding terms are found from their two predecessors by the equations

$$x_3 = ax_2 + bx_1$$

$$x_4 = ax_3 + bx_2$$

$$x_5 = ax_4 + bx_3$$

and, indeed, for each positive integer n bigger than 2,

$$x_n = ax_{n-1} + bx_{n-2}.$$

One very well-known example of a second-order recursion is the *Fibonacci sequence*:

$$\{1, 1, 2, 3, 5, 8, 13, 21, 34, 55, 89, 144, 233, 377, 610, 987, 1597, \ldots\}.$$

Here, the multipliers are both equal to 1 and each term from the third on is the sum of the two terms before it. That is, for each positive integer n exceeding 2:

$$x_n = x_{n-1} + x_{n-2}.$$

The Fibonacci sequence is such a storehouse of interesting mathematical relationships that there is a special journal, *The Fibonacci Quarterly*, devoted to its study. We will look at some of these that involve powers of numbers.

The sums of the squares of two adjacent entries yields alternate terms in the sequence:

$$1^2 + 1^2 = 2; \quad 1^2 + 2^2 = 5; \quad 2^2 + 3^2 = 13;$$

$$3^2 + 5^2 = 34; \quad 5^2 + 8^2 = 89; \quad \ldots$$

The remaining terms are differences of squares of other terms:

$$2^2 - 1^2 = 3; \quad 3^2 - 1^2 = 8; \quad 5^2 - 2^2 = 21;$$

$$8^2 - 3^2 = 55; \quad 13^2 - 5^2 = 144; \quad \ldots$$

The square of any term differs from the product of its two neighbors by 1:

$$3^2 = 2 \times 5 - 1; \quad 5^2 = 3 \times 8 + 1;$$
$$8^2 = 5 \times 13 - 1; \quad 13^2 = 5 \times 34 - 1; \ \ldots$$

Three successive cubes of terms in the sequence can be used to yield later terms:

$$2^3 + 1^3 - 1^3 = 8$$
$$3^3 + 2^3 - 1^3 = 34$$
$$5^3 + 3^3 - 2^3 = 144$$
$$8^3 + 5^3 - 3^3 = 610.$$

You can check these relationships out as far along in the sequence as you wish. This, in itself, does not prove that they will *always* occur beyond where you have checked. However, mathematics is not merely an empirical science. We can establish by logical argument that the relationships we have indicated will hold in general. What is more, any student completing high school mathematics should know enough mathematics to understand and perhaps generate the proofs (4).

Going back to squares, let us look at the squares of the entries of the Fibonacci sequence and take a running total:

$$1^2 + 1^2 = 1 \times 2$$
$$1^2 + 1^2 + 2^2 = 2 \times 3$$
$$1^2 + 1^2 + 2^2 + 3^2 = 3 \times 5$$
$$1^2 + 1^2 + 2^2 + 3^2 + 5^2 = 5 \times 8$$
$$1^2 + 1^2 + 2^2 + 3^2 + 5^2 + 8^2 = 8 \times 13.$$

By choosing the initial terms differently, we can construct other sequences which satisfy the recursion relation $x_{n+1} = x_n + x_{n-1}$. For example, the Lucas sequence $\{1, 3, 4, 7, 11, 18, 29, 47, \ldots\}$ has been well studied. For each such sequence,

$$(x_n x_{n+3}, 2x_{n+1}x_{n+2}, x_{n+1}^2 + x_{n+2}^2)$$

turns out to be always a pythagorean triple. The Fibonacci sequence $\{1, 1, 2, 3, \ldots\}$ yields the triples $(3, 4, 5), (5, 12, 13), (16, 30, 34), (39, 80, 89)$; the Lucas sequence yields the triples $(7, 24, 25), (33, 56, 65)$; the sequence $\{3, 1, 4, 5, 9, 14, \ldots\}$ yields the triples $(15, 8, 17), (9, 40, 41), (56, 90, 106)$ (5).

There are other second-order recursions which have interesting properties with respect to squares. We have already met some of them in the previous section:

$$1, 6, 35, 204, 1189, 6930, 40391, \ldots$$
$$1, 2, 5, 12, 29, 70, 169, \ldots$$
$$1, 3, 7, 17, 41, 99, 239, \ldots \ .$$

More generally, fix an integer k and define

$$u_1 = 1, u_2 = k, u_n = ku_{n-1} - u_{n-2} \quad \text{for } n \geq 3$$

and

$$v_1 = 1, v_2 = k, v_n = kv_{n-1} + v_{n-2} \quad \text{for } n \geq 3.$$

A few examples are

k	$\{u_n\}$	$\{v_n\}$
2	$\{1, 2, 3, 4, 5, \ldots\}$	$\{1, 2, 5, 12, 29, \ldots\}$
3	$\{1, 3, 8, 21, 55, \ldots\}$	$\{1, 3, 10, 33, 109, \ldots\}$
4	$\{1, 4, 15, 56, 209, \ldots\}$	$\{1, 4, 17, 72, 305, \ldots\}$
5	$\{1, 5, 24, 115, 551, \ldots\}$	$\{1, 5, 26, 135, 701, \ldots\}$

We have the identities **(6)**

$$u_n^2 = u_{n+1}u_{n-1} + 1$$
$$u_{n+1}^2 - u_n^2 = u_{2n+1}$$
$$v_n^2 = v_{n+1}v_{n-1} + (-1)^{n-1}$$
$$v_{n+1}^2 + v_n^2 = v_{2n+1}.$$

Second-order recursions have an interesting property which is related to the magic square:

$$
\begin{array}{ccc}
4 & 3 & 8 \\
9 & 5 & 1 \\
2 & 7 & 6.
\end{array}
$$

Each row, each column, and each diagonal consists of three numbers which add up to 15. Look at the products of the terms in each row and of the terms in each column. We find that the sum of the row products is equal to the sum of the column products:

$$4 \times 3 \times 8 + 9 \times 5 \times 1 + 2 \times 7 \times 6 = 96 + 45 + 84 = 225 = 15^2$$

and

$$4 \times 9 \times 2 + 3 \times 5 \times 7 + 8 \times 1 \times 6 = 72 + 105 + 48 = 225 = 15^2.$$

The common sum is the square of the sum of any row, column or diagonal.

Now, take any nine consecutive terms of a second-order recursion

$$x_1, x_2, x_3, x_4, x_5, x_6, x_7, x_8, x_9$$

and put them into a rectangular array so that the indices of the xs form a magic square:

$$
\begin{array}{ccc}
x_4 & x_3 & x_8 \\
x_9 & x_5 & x_1 \\
x_2 & x_7 & x_6
\end{array}
$$

As for the magic square, form the sum of the products of the terms in the three rows and the sum of the products of the terms in the three columns. These sums will always be equal (7):

$$x_4 x_3 x_8 + x_9 x_5 x_1 + x_2 x_7 x_6 = x_4 x_9 x_2 + x_3 x_5 x_7 + x_8 x_1 x_6.$$

Unlike the situation for the magic square, the common sum will not in general be square. However, if we try this on the first nine terms of the Fibonacci sequence, we get the square array:

$$
\begin{array}{ccc}
3 & 2 & 21 \\
34 & 5 & 1 \\
1 & 13 & 8.
\end{array}
$$

The sum of the row products is $3 \times 2 \times 21 + 34 \times 5 \times 1 + 1 \times 13 \times 8 = 126 + 170 + 104 = 400 = 20^2$ and the sum of the column products is $102 + 130 + 168 = 400$.

A case of a second-order nonlinear recursion that surprisingly generates only squares is given by the following description (8). Suppose a, b, c, d are any integers. Define

$$x_0 = a^2, \qquad x_1 = b^2$$

$$x_{n+1} = (c^2 - 2d)x_n - d^2 x_{n-1} + 2(b^2 + a^2 d - abc)d^n \quad \text{for } n \geq 1.$$

Examples are

$$
\begin{array}{ll}
\{1, 4, 1, 1, 4, 1, 1, \ldots\} & (a, b, c, d) = (1, 2, 1, 1) \\
\{1, 4, 16, 64, 256, \ldots\} & (a, b, c, d) = (1, 2, 3, 2) \\
\{4, 9, 64, 676, 7744, \ldots\} & (a, b, c, d) = (2, 3, 4, 2).
\end{array}
$$

Catalan numbers

One sequence where squares arise is that of *Catalan* numbers **(9)**:

$$1, 1, 2, 5, 14, 42, 132, 429, \ldots.$$

The nth Catalan number is the number of ways a string of n letters in fixed order can be bracketed with $n - 1$ pairs of parentheses with two terms between each pair. Thus, when $n = 4$, we have five possibilities: $((ab)(cd))$; $(((ab)c)d)$; $(a(b(cd)))$; $(a((bc)d))$; $((a(bc))d)$. Alternatively, the nth number is the number of ways a regular polygon with $n + 1$ sides can be decomposed into triangles by nonintersecting diagonals.

The law of formation for the Catalan numbers is a little complicated. We start off with two ones. Each entry after that is the sum of products of pairs of the previous terms, working in from both ends. For example,

$$429 = 1 \times 132 + 1 \times 42 + 2 \times 14 + 5 \times 5 + 14 \times 2 + 42 \times 1 + 132 \times 1.$$

If we denote the successive terms of the Catalan sequence by $x_1, x_2, \ldots,$ $x_n, \ldots$, we find that

$$1 + 24 \left[\frac{x_{n-1} x_{n-2}}{x_n x_{n-2} - x_{n-1}^2} \right]$$

runs through consecutive squares of odd numbers. For example

$$1 + 24 \left[\frac{5 \times 14}{5 \times 42 - 14^2} \right] = 1 + 24 \left[\frac{70}{14} \right] = 121 = 11^2.$$

Catalan numbers and squares are related also through the Pascal triangle

```
                    1
                1       1
            1       2       1
        1       3       3       1
    1       4       6       4       1
1       5       10      10      5       1
```

The first rows are as given. The end elements of each row are always 1. Each interior element is the sum of the two closest elements in the row above. For the rows, the sum of the squares of the elements are respectively $1 = 1 \times 1$, $2 = 2 \times 1$, $6 = 3 \times 2$, $20 = 4 \times 5$, $70 = 5 \times 14$, $252 = 6 \times 42, \ldots,$ i.e., products of natural numbers and Catalan numbers. These sums occur also as the middle entries in the odd-numbered rows.

Pascal's triangle, the Catalan numbers and powers of two are intertwined in a further way. We form products of corresponding consecutive terms of the three sequences and sum with alternating sign. The tables below will illustrate:

1	2	1
1	2	5
4	2	1
4	8	5

$$4 - 8 + 5 = 1.$$

1	3	3	1
1	2	5	14
8	4	2	1
8	24	30	14

$$8 - 24 + 30 - 14 = 0.$$

1	4	6	4	1
1	2	5	14	42
16	8	4	2	1
16	64	120	112	42

$$16 - 64 + 120 - 112 + 42 = 2.$$

1	5	10	10	5	1
1	2	5	14	42	132
32	16	8	4	2	1
32	160	400	560	420	132

$$32 - 160 + 400 - 560 + 420 - 132 = 0.$$

The reader is invited to explore this one further **(10)**.

Sequences of polynomial values

The simplest type of sequence is an arithmetic progression, in which the difference between each pair of consecutive terms remains unchanged. For example, 1, 25, 49 are three squares in arithmetic progression, the common difference between consecutive terms being 24. The triples 49, 169, 289 and 49, 289, 529 each also consist of squares in arithmetic progression, and we find that 49, 169, 289, 409, 529 is an arithmetic progression of five terms, four of which are perfect squares.

Is it possible to find more than three squares in a row in arithmetic progression? The answer to this, as Euler discovered two hundred years ago, is no **(11)**.

An arithmetic progression consists of terms of the form $at+b$ where a and b are fixed integers, and t runs through consecutive integers. This suggests that we might ask about nonlinear polynomials $p(t)$ which take square values for a succession of integer values of t. For example, if $p(t) = at^2 + bt + c$, $p(t)$ will always assume square values if $p(t)$ is the square of a linear polynomial. But how long a run of squares can we get for a quadratic polynomial otherwise?

Here are some quadratics along with the consecutive values of t for which their values are square:

$$60t^2 - 60t + 1 \qquad t = -2, -1, 0, 1, 2, 3$$
$$-4980t^2 + 32100t + 2809 \qquad t = 0, 1, 2, 3, 4, 5, 6$$
$$-420t^2 + 2100t + 2809 \qquad t = -1, 0, 1, 2, 3, 4, 5, 6$$

Can the reader find a quadratic which is square for a longer stretch of consecutive values of t but which is not the square of a linear polynomial?

How long a stretch of square values is possible for a cubic polynomial or a quartic polynomial not the square of a quadratic? One interesting quartic polynomial *is* the square of a quadratic: $t(t+1)(t+2)(t+3)+1 = (t^2+3t+1)^2$ **(12)**. This says that the product of any four consecutive integers is always one less than a perfect square. In particular, $4! = 24 = 5^2 - 1$. There are two other factorials that are one less than a perfect square, $5!$ and $7!$. Moreover, the numbers $720 = 6!$ and $5040 = 7!$ both have the property that they differ from the next three greater squares by squares:

$$720 = 27^2 - 3^2 = 28^2 - 8^2 = 29^2 - 11^2$$
$$5040 = 71^2 - 1^2 = 72^2 - 12^2 = 73^2 - 17^2.$$

Are there any numbers that differ from the next *four* greater squares by squares? For such a number N, there must exist integers z, a, b, c for which

$$(z-1)^2 \le N = z^2 - a^2 = (z+1)^2 - b^2 = (z+2)^2 - c^2 = (z+3)^2 - d^2.$$

Eliminating z and N from this chain of equations yields

$$2(b^2 + 1) = a^2 + c^2 \qquad\qquad\text{(i)}$$
$$2(c^2 + 1) = b^2 + d^2 \qquad\qquad\text{(ii)}$$
$$b^2 \ge 2(a^2 + 1). \qquad\qquad\text{(iii)}$$

The third of these is a consequence of $2z - 1 \ge a^2$ and $2z + 1 = b^2 - a^2$. The first two equations assert that $b^2 - a^2$, $c^2 - b^2$ and $d^2 - b^2$ are in arithmetic progression.

A computer search leads to a variety of solutions for (i) and (ii) including $(a, b, c, d) = (6, 23, 32, 39)$, $(39, 70, 91, 108)$, $(108, 157, 194, 225)$,

$(225, 296, 353, 402)$. The solutions fall into a regular sequence $\{a_n, b_n, c_n, d_n\}$ where

$$a_n = 2n^3 - 5n = 2(n + \tfrac{1}{2})^3 - 3(n + \tfrac{1}{2})^2 - \tfrac{7}{2}(n + \tfrac{1}{2}) + \tfrac{9}{4}$$

$$b_n = 2n^3 + 2n^2 - n + 1 = 2(n + \tfrac{1}{2})^3 - (n + \tfrac{1}{2})^2 - \tfrac{3}{2}(n + \tfrac{1}{2}) + \tfrac{7}{4}$$

$$c_n = 2n^3 + 4n^2 + n - 2 = 2(n + \tfrac{1}{2})^3 + (n + \tfrac{1}{2})^2 - \tfrac{3}{2}(n + \tfrac{1}{2}) - \tfrac{7}{4}$$

$$d_n = 2n^3 + 6n^2 + n - 3 = 2(n + \tfrac{1}{2})^3 + 3(n + \tfrac{1}{2})^2 - \tfrac{7}{2}(n + \tfrac{1}{2}) - \tfrac{9}{4},$$

and $d_n = a_{n+1}$. The solutions given belong to $2 \leq n \leq 5$, and the first three of them also satisfy (iii), yielding

$$245^2 < 60480 = 246^2 - 6^2 = 247^2 - 23^2 = 248^2 - 32^2$$

$$= 249^2 - 39^2$$

$$1688^2 < 2851200 = 1689^2 - 39^2 = 1690^2 - 70^2 = 1691^2 - 91^2$$

$$= 1692^2 - 108^2$$

$$6491^2 < 42134400 = 6492^2 - 108^2 = 6493^2 \quad 157^2 = 6494^2 - 194^2$$

$$= 6495^2 - 225^2.$$

It is an open question as to whether there is a number that differs from the next five greater squares by a square **(13)**.

Exercises on the Notes

1. *High school; basic properties of numbers; moderate difficulty* Suppose we form a sequence $\{p_n\}$ of primes as follows. Let $p_1 = 2$. Suppose that $p_1, p_2, \ldots, p_{n-1}$ have already been chosen. Let p_n be the *largest* prime factor of $1 + p_1 p_2 \ldots p_{n-1}$. Thus $\{p_n\} = \{2, 3, 7, 43, 139, \ldots\}$. Can p_n ever be 5?

2. *Simple algebra.* Suppose that we start with the pythagorean triple $(2k+1, 2k^2+2k, 2k^2+2k+1)$. Find the sequences obtained for the difference and the ratio of the middle terms of successive triples.

3. *Pattern recognition; moderately difficult.* Indicate how to obtain an infinite sequence of pythagorean triples whose smallest terms have a difference of 17.

4. *Algebra; moderate to difficult.* Let $f_0 = 0$ $f_1 = f_2 = 1$ and $f_{n+1} = f_n + f_{n-1}$ for $n \geq 1$.

(a) Suppose that $u_n = f_n^2 + f_{n+1}^2$ and $v_n = f_{2n+1}$. Prove that $u_0 = v_0$, $u_1 = v_1$, and that $u_{n+1} = 3u_n - u_{n-1}$ and $v_{n+1} = 3v_n - v_{n-1}$ for $n \geq 1$; use this to show that $f_n^2 + f_{n+1}^2 = f_{2n+1}$ for $n \geq 0$.

(b) By writing f_{2n} as a difference of its neighbors, or otherwise, prove that $f_{2n} = f_{n+1}^2 - f_{n-1}^2$ for $n \geq 1$.

(c) Prove by induction that $f_n^2 = f_{n-1}f_{n+1} + (-1)^{n-1}$ for $n \geq 1$.

(d) Suppose that $g_n = f_{n+1}^3 + f_n^3 - f_{n-1}^3$ and $h_n = f_{3n}$. Prove that $g_{n+1} = 4g_n + g_{n-1}$ and $h_{n+1} = 4h_n + h_{n-1}$ for $n \geq 2$, and deduce that $f_{n+1}^3 + f_n^3 - f_{n-1}^3 = f_{3n}$ for $n \geq 1$.

(e) Prove by induction that $f_0^2 + f_1^2 + \cdots + f_n^2 = f_n f_{n+1}$ for $n \geq 0$.

5. *Simple algebra.* If $x_{n+1} = x_n + x_{n-1}$ for each subscript n, verify that $(x_n x_{n+3}, 2x_{n+1}x_{n+2}, x_{n+1}^2 + x_{n+2}^2)$ is a pythagorean triple.

6. *Moderate algebra.* Establish the four identities, as well as the identities $u_n(u_{n+1} - u_{n-1}) = u_{2n}$ and $v_n(v_{n+1} + v_{n-1}) = v_{2n}$.

7. *College level elementary linear algebra; moderately difficult.* Let a and b be complex parameters and define

$$V = \{\mathbf{x} = (x_0, x_1, x_2, \ldots, x_n, \ldots) : x_n = ax_{n-1} + bx_{n-2} \text{ for } n \geq 2\}$$

be a complex vector space of infinite complex sequences with multiplication by scalars and addition defined coordinatewise:

$$c(x_0, x_1, x_2, \ldots, x_n, \ldots) = (cx_0, cx_1, cx_2, \ldots, cx_n, \ldots)$$

$$(x_0, x_1, x_2, \ldots, x_n, \ldots) + (y_0, y_1, y_2, \ldots, y_n, \ldots)$$

$$= (x_0 + y_0, x_1 + y_1, x_2 + y_2, \ldots, x_n + y_n, \ldots).$$

(a) Prove that V has a basis consisting of the two vectors

$$(1, 0, b, ab, a^2 b + b^2, a^3 b + 2ab^2, \ldots)$$

and

$$(0, 1, a, a^2 + b, a^3 + 2ab, \ldots)$$

and so V has dimension 2.

(b) Prove that the geometric sequence $(1, r, r^2, r^3, \ldots, r^n, \ldots)$ belongs to V if and only if $r^2 = ar + b$.

(c) Suppose that the equation $t^2 = at + b$ has distinct roots λ and μ. Prove that $\mathbf{\Lambda} = (1, \lambda, \lambda^2, \ldots, \lambda^n, \ldots)$ and $\mathbf{M} = (1, \mu, \mu^2, \mu^3, \ldots, \mu^n, \ldots)$ constitute a linearly independent pair of vectors, and thus form a basis for V.

Deduce that, given any element $\mathbf{x}$ in V, there exist scalars α and β for which $\mathbf{x} = \alpha\mathbf{\Lambda} + \beta\mathbf{M}$, i.e., $x_n = \alpha\lambda^n + \beta\mu^n$ for each nonnegative integer n.

(d) Suppose that $t^2 = at + b$ has coincident roots λ, so that $a^2 = -4b$ and $\lambda = a/2$. Prove that

$$\mathbf{\Lambda} = (1, \lambda, \lambda^2, \ldots, \lambda^n, \ldots) \quad \text{and} \quad \mathbf{\Lambda}' = (0, \lambda, 2\lambda^2, 3\lambda^3, \ldots, n\lambda^n, \ldots)$$

constitute a basis for V and deduce that, for any vector $\mathbf{x}$, there exist scalars α and β for which $\mathbf{x} = \alpha\mathbf{\Lambda} + \beta\mathbf{\Lambda}'$, i.e., $x_n = (\alpha + \beta n)\lambda^n$ for each nonnegative integer n.

(e) Let

$$f_0 = 0, f_1 = f_2 = 1, f_n = f_{n-1} + f_{n-2} \quad (n \geq 2)$$

$$u_0 = 0, u_1 = 1, u_2 = k, u_n = ku_{n-1} - u_{n-2} \quad (n \geq 2)$$

$$v_0 = 0, v_1 = 1, v_2 = k, v_n = kv_{n-1} + v_{n-2} \quad (n \geq 2).$$

Give a general formula for the nth term of each sequence. In particular, show that, when $0 < k < 2$,

$$u_n = \frac{2 \sin n\theta}{\sqrt{4 - k^2}}$$

for $\sin \theta = \frac{1}{2}\sqrt{4 - k^2}$ and $\cos \theta = \frac{k}{2}$, while, for $k > 2$,

$$u_n = \frac{2 \sinh n\psi}{\sqrt{k^2 - 4}}$$

where $\sinh \psi = \frac{1}{2}\sqrt{k^2 - 4}$ and $\cosh \psi = \frac{k}{2}$. [HINT: The roots of $t^2 = kt - 1$ are $\cos \theta \pm \sin \theta$ when $0 < k < 2$ and $\cosh \psi \pm \sinh \psi$ when $k > 2$.]

(f) Using (c) and (d), or otherwise, prove that for any second-order linear recursion

$$x_4x_3x_8 + x_9x_5x_1 + x_2x_7x_6 = x_4x_9x_2 + x_3x_5x_7 + x_8x_1x_6.$$

Notes

1. See the remarks on page 4 of Paulo Ribenboim, *The book of prime number records*, *Second edition*, Springer-Verlag, 1984.

C. D. Cox and A. J. Van der Poorten, in *Australian Mathematical Society Journal* 8 (1968), 571–574 show that if we start with the prime 2 and the largest possible prime factor is selected each time, then infinitely many primes including 5, 11, 13, 17, and 19 will not be found.

The following problem appears to be unsolved: Suppose that P is a set of at least two primes such that for all p and q in P, all the prime divisors of $pq + 1$ are in P. Is P the set of all primes? (*Mathematics Magazine* 54 (1981), 84; 55 (1982), 180)

In *American Mathematical Monthly* 96 (1989), 339–341, it is shown that, if $p_1, p_2, \ldots, p_n$ are the first n primes, then $p_1 \cdots p_{i-1} p_{i+1} \cdots p_n + 1$ cannot be a power of p_i when $n \geq 4$.

2. In each case, we move from a triple $(2k + 1, 2k^2 + 2k, 2k^2 + 2k + 1)$ to

$$(2k^2 + 2k + 1, 2(k^2 + k)^2 + 2(k^2 + k), 2(k^2 + k)^2 + 2(k^2 + k) + 1)$$

$$= (2k^2 + 2k + 1, (2k^2 + 2k)(k^2 + k + 1), (2k^2 + 2k)(k^2 + k + 1) + 1).$$

The ratio of successive middle terms is $k^2 + k + 1$ and their difference is $2k^2(k + 1)^2$. Suppose that k_m is the value of the parameter k for the mth triple in the sequence. Then $k_{m+1} = k_m(k_m + 1)$ for $m \geq 1$. The ratio and difference of the middle terms are respectively $k_{m+1} + 1$ and $2k_{m+1}^2$. Note that

$$k_{m+1} = k_m(k_m + 1) = k_1(k_1 + 1)(k_2 + 1) \cdots (k_m + 1) \quad (m \geq 1).$$

3. The generation of pythagorean triples whose smallest two numbers differ by a fixed amount can be put in a more general setting. Define the sequence $\{t_n\}$ by the second-order recursion

$$t_0 = a, \qquad t_1 = b, \qquad t_{n+1} = ut_n - t_{n-1},$$

where a, b, and u are given parameters. For example, the choice $a = 0$, $b = 1$, and $u = 6$ yields the sequence $\{0, 1, 6, 35, 204, \ldots\}$. Suppose that the numbers r, s, k satisfy $r + s = k$, $k = \frac{u+2}{u-2}$, that d is a fixed "difference" and that the system

$$rx + sy = w$$

$$rx^2 + sy^2 = z^2$$

$$y - x = d$$

has a solution $(x, y, z, w) = (x_0, y_0, z_0, w_0)$ with $z_0 = b - a$ and $w_0 = b + a$. For $n \geq 1$, define

$$z_n = t_{n+1} - t_n \qquad w_n = t_{n+1} + t_n$$

$$x_n = x_{n-1} + \frac{ut_n - 2t_{n-1}}{k} = x_{n-1} + \frac{t_{n+1} - t_{n-1}}{k}$$

$$y_n = x_n + d.$$

Then it can be shown by induction that $(x, y, z, w) = (x_n, y_n, z_n, w_n)$ satisfies the foregoing system for $n \geq 0$. This is given as true for $n = 0$. Suppose that

$$rx_{n-1} + sy_{n-1} = w_{n-1}, \quad rx_{n-1}^2 + sy_{n-1}^2 = z_{n-1}^2, \quad y_{n-1} = x_{n-1} + d.$$

From the definition, $y_n = x_n + d$ and, straightforwardly,

$$rx_n + sy_n = kx_n + sd = kx_{n-1} + sd + (t_{n+1} - t_{n-1})$$
$$= w_{n-1} + (t_{n+1} - t_{n-1})$$
$$= (t_n + t_{n-1}) + (t_{n+1} - t_{n-1}) = w_n.$$

Also

$$rx_n^2 + sy_n^2 = (r + s)x_n^2 + 2sdx_n + sd^2$$
$$= kx_{n-1}^2 + 2k\left(\frac{ut_n - 2t_{n-1}}{k}\right)x_{n-1} + k\left(\frac{ut_n - 2t_{n-1}}{k}\right)^2$$
$$\quad + 2sdx_{n-1} + 2s\left(\frac{ut_n - 2t_{n-1}}{k}\right)d + sd^2$$
$$= (kx_{n-1}^2 + 2sdx_{n-1} + sd^2)$$
$$\quad + \tfrac{2}{k}(ut_n - 2t_{n-1})(kx_{n-1} + sd) + \tfrac{1}{k}(ut_n - 2t_{n-1})^2$$
$$= z_{n-1}^2 + \tfrac{2}{k}(ut_n - 2t_{n-1})w_{n-1} + \tfrac{1}{k}(ut_n - 2t_{n-1})^2$$
$$= \tfrac{1}{k}\big[k(t_n - t_{n-1})^2 + 2(ut_n - 2t_{n-1})(t_n + t_{n-1})$$
$$\quad + (ut_n - 2t_{n-1})^2\big]$$
$$= \tfrac{1}{k}\big[(k + 2u + u^2)t_n^2 + 2(-k + u - 2 - 2u)t_n t_{n-1}$$
$$\quad + (k - 4 + 4)t_{n-1}^2\big].$$

Since $u + 2 = k(u - 2)$, this expression reduces to

$$(1 + u^2 - 2u)t_n^2 + 2(1 - u)t_n t_{n-1} + t_{n-1}^2 = \big[(u - 1)t_n - t_{n-1}\big]^2$$
$$= (t_{n+1} - t_n)^2 = z_n^2$$

as desired.

For example, suppose we wish to determine pythagorean triples (x, y, z) for which $y - x = 7$. Take $r = s = 1, k = 2, u = 6$. One possibility is $(x, y, z) = (-4, 3, 5)$, so we select the initial terms $t_0 = a, t_1 = b$ to satisfy $b + a = (-4) + 3 = -1$ and $b - a = \sqrt{(-4)^2 + 3^2} = 5$, i.e., $t_0 = -3$ and $t_1 = 2$. Embed these terms in a recursion satisfying $t_{n+1} = 6t_n - t_{n-1}$ for integer indices n. This yields the sequence $\{\ldots, -117, -20, -3, 2, 15, 88, \ldots\}$ and the triples

$$\{\ldots, (-72, -65, 97), (-15, -8, 17), (-4, 3, 5), (5, 12, 13), (48, 55, 73), \ldots\}.$$

More generally, given the pythagorean triple $(m^2 - n^2, 2mn, m^2 + n^2)$, we can obtain a parametric family of others with the same difference $2mn - (m^2 - n^2)$ between the lowest terms by taking $r = s = 1$, $k = 2$, $u = 6$, $t_0 = mn - n^2$, $t_1 = m^2 + mn$. Some pythagorean triples quadratic in m and n obtained are $(4m^2 + 2mn, 3m^2 + 4mn + n^2, 5m^2 + 4mn + n^2)$ and $(21m^2 + 16mn + 3n^2, 20m^2 + 18mn + 4n^2, 29m^2 + 24mn + 5n^2)$.

A given sequence, even of integers, satisfying the recursion $t_{n+1} = 6t_n - t_{n-1}$ need not correspond to integer pythagorean triples. For example, $\{\ldots, 1, 1, 5, 29, 169, \ldots\}$ generates triples with difference $2i = 2\sqrt{-1}$, including $(1-i, 1+i, 0)$, $(3-i, 3+i, 4)$, $(17-i, 17+i, 24)$, $(99-i, 99+i, 140)$. Readers will note the intervention of the sequences $\{\ldots, 0, 1, 2, 5, 12, 29, \ldots\}$ and $\{\ldots, 1, 1, 3, 7, 17, 41, \ldots\}$ whose terms are factors of the terms in the sequence $\{\ldots, 0, 1, 6, 35, 204, \ldots\}$.

Consider the diophantine equation $x^2 + 2y^2 = z^2$. To get a family of solutions that fits within this framework, we want to take $r = 1$, $s = 2$, $k = 3$, so that u should equal 4. For the sequence $\{1, 4, 15, 56, 209, \ldots\}$ mentioned in the text, we form the table:

z_n	t_n	w_n	equations
	1		
3		$5 = 1 + 2 \cdot 2$	$1^2 + 2 \cdot 2^2 = 3^2$
	4		
11		$19 = 7 + 2 \cdot 6$	$7^2 + 2 \cdot 6^2 = 11^2$
	15		
41		$71 = 23 + 2 \cdot 24$	$23^2 + 2 \cdot 24^2 = 41^2$
	56		
153		$265 = 89 + 2 \cdot 88$	$89^2 + 2 \cdot 88^2 = 153^2$
	209		

A superficial glance suggests that this is all according to pattern. But wait a minute! The differences between x and y alternate in sign: $2 - 1 = +1$; $6 - 7 = -1$; $24 - 23 = +1$; $88 - 89 = -1$. If we take account of the parametrization introduced earlier on, using both the values $d = 1$ and $d = -1$, we find that our sequence of solutions to $x^2 + 2y^2 = z^2$ is actually an interlacing of the integer members of two sequences of solutions:

$$(1, 2, 3), (17/3, 20/3, 11), (23, 24, 41), (263/3, 266/3, 153), \ldots$$

$$(7/3, 4/3, 3), (7, 6, 11), (73/3, 70/3, 41), (89, 88, 153), \ldots.$$

In fact, the process described will yield rational rather than integer solutions to $rx^2 + sy^2 = z^2$ in general, although we can get integer solutions by multiplying through by a common denominator. For example, starting with the solution $(x, y, z) = (1, 4, 7)$ to $x^2 + 3y^2 = z^2$, we are led to consider the sequence $\{3, 10, 91/3, \ldots\}$ where $k = 4, d = 3$ and $u = 10/3$. The next solution of $x^2 + 3y^2 = z^2$ is $(47/6, 65/6, 61/3)$, which multiplies up to the integer solution $(47, 65, 122)$.

We can adapt the method to the situation that the coefficient of z^2 differs from 1. The equation $3x^2 + 2y^2 = 2z^2$ has solution $(4, 5, 7)$. Rewrite the equation as $(3/2)x^2 + y^2 = z^2$, and take $k = 5/2$, $u = 14/3$. The relevant sequence is $\{2, 9, 40, \ldots\}$. The next solution of the diophantine equation is $(96/5, 101/5, 31)$, which yields the integer solution $(96, 101, 155)$.

4. Let $f_0 = 0$, $f_1 = f_2 = 1$ and $f_{n+1} = f_n + f_{n-1}$ for $n \geq 1$ define the Fibonacci sequence. We need to establish that

(a) $f_n^2 + f_{n+1}^2 = f_{2n+1}$ for $n \geq 0$;

(b) $f_{n+1}^2 - f_{n-1}^2 = f_{2n}$ for $n \geq 1$;

(c) $f_n^2 = f_{n-1}f_{n+1} + (-1)^{n-1}$ for $n \geq 1$;

(d) $f_{n+1}^3 + f_n^3 - f_{n-1}^3 = f_{3n}$ for $n \geq 1$; and

(e) $f_0^2 + f_1^2 + \cdots + f_n^2 = f_n f_{n+1}$ for $n \geq 0$.

Ad (a): Let $u_n = f_n^2 + f_{n+1}^2$ and $v_n = f_{2n+1}$. We prove (a) by establishing that $u_0 = v_0$, $u_1 = v_1$, $u_{n+1} = 3u_n - u_{n-1}$ and $v_{n+1} = 3v_n - v_{n-1}$ for $n \geq 1$. An induction argument will then show that $u_n = v_n$ for each n. That $u_0 = v_0$ and $u_1 = v_1$ is easily checked.

$$u_{n+1} - 3u_n + u_{n-1} = f_{n+2}^2 - 2f_{n+1}^2 - 2f_n^2 + f_{n-1}^2$$
$$= (f_{n+1} + f_n)^2 - 2f_{n+1}^2 - 2f_n^2 + (f_{n+1} - f_n)^2$$
$$= 0;$$

and

$$v_{n+1} - 3v_n + v_{n-1} = f_{2n+3} - 3f_{2n+1} + f_{2n-1}$$
$$= (f_{2n+3} - f_{2n+1}) - f_{2n+1} - (f_{2n+1} - f_{2n-1})$$
$$= f_{2n+2} - f_{2n+1} - f_{2n} = 0.$$

Ad (b):

$$f_{2n} = f_{2n+1} - f_{2n-1}$$
$$= (f_n^2 + f_{n+1}^2) - (f_{n-1}^2 + f_n^2) = f_{n+1}^2 - f_{n-1}^2.$$

Ad (c): This follows from the fact that, for $n \geq 1$,

$$f_{n+1}^2 - f_{n+2}f_n = f_{n+1}^2 - (f_{n+1} + f_n)f_n$$
$$= f_{n+1}(f_{n+1} - f_n) - f_n^2 = f_{n+1}f_{n-1} - f_n^2.$$

Ad (d): By inspection, we see that the successive values of both sides of $f_{n+1}^3 + f_n^3 - f_{n-1}^3 = f_{3n}$ are 2, 8, 34, 144, 610, $\ldots$, which satisfies the recursion relation $x_{n+1} = 4x_n + x_{n-1}$. Accordingly, define

$$g_n = f_{n+1}^3 + f_n^3 - f_{n-1}^3$$

and $h_n = f_{3n}$. We will establish (c) by showing that $g_1 = h_1$, $g_2 = h_2$, $g_{n+1} = 4g_n + g_{n-1}$ and $h_{n+1} = 4h_n + h_{n-1}$ for $n \geq 2$.

$$g_{n+1} - 4g_n - g_{n-1}$$

$$= f_{n+2}^3 - 3f_{n+1}^3 - 6f_n^3 + 3f_{n-1}^3 + f_{n-2}^3$$

$$= (f_{n+2}^3 - f_{n+1}^3) - 2(f_{n+1}^3 - f_{n-1}^3) - 6f_n^3 + (f_{n-1}^3 + f_{n-2}^3)$$

$$= f_n [f_{n+2}^2 + f_{n+2}f_{n+1} + f_{n+1}^2 - 2(f_{n+1}^2 + f_{n+1}f_{n-1} + f_{n-1}^2)$$

$$\quad - 6f_n^2 + (f_{n-1}^2 - f_{n-1}f_{n-2} + f_{n-2}^2)]$$

$$= f_n [f_{n+1}^2 + 2f_{n+1}f_n + f_n^2 + f_{n+1}^2 + f_{n+1}f_n + f_{n+1}^2$$

$$\quad - 2(f_{n+1}^2 + f_{n+1}f_{n-1} + f_{n-1}^2) - 6f_n^2$$

$$\quad + (f_{n-1}^2 - f_nf_{n-1} + f_{n-1}^2 + f_n^2 - 2f_nf_{n-1} + f_{n-1}^2)]$$

$$= f_n(f_{n+1}^2 + 3f_{n+1}f_n - 4f_n^2 - 2f_{n+1}f_{n-1} - 3f_nf_{n-1} + f_{n-1}^2)$$

$$= f_n(f_{n+1} + 4f_n - f_{n-1})(f_{n+1} - f_n - f_{n-1}) = 0.$$

$$h_{n+1} - 4h_n - h_{n-1} = f_{3n+3} - 4f_{3n} - f_{3n-3}$$

$$= f_{3n+2} + f_{3n+1} - 4f_{3n} - f_{3n-1} + f_{3n-2}$$

$$= (f_{3n+1} + f_{3n}) + f_{3n+1} - 4f_{3n} - f_{3n-1}$$

$$\quad + (f_{3n} - f_{3n-1})$$

$$= 2(f_{3n+1} - f_{3n} - f_{3n-1}) = 0.$$

An alternative proof uses matrices. Let

$$Q = \begin{pmatrix} 1 & 1 \\ 1 & 0 \end{pmatrix}.$$

Then

$$Q^n = \begin{pmatrix} f_{n+1} & f_n \\ f_n & f_{n-1} \end{pmatrix}.$$

Then comparison of the corresponding off diagonal entries of the two sides of the equation $Q^{3n} = (Q^n)^3$ yields the result.

Ad (e): Observe that $f_nf_{n+1} + f_{n+1}^2 = f_{n+1}(f_n + f_{n+1}) = f_{n+1}f_{n+2}$.

5. This was pointed out in a letter by Herta T. Freitag in the *Mathematics Teacher* 85 (Jan., 1992), 10.

6. Extend both sequences by setting $u_0 = v_0 = 0$. Since

$$u_{n+1}^2 - u_{n+2}u_n = u_{n+1}^2 - (ku_{n+1} - u_n)u_n$$
$$= u_{n+1}(u_{n+1} - ku_n) + u_n^2 = u_n^2 - u_{n+1}u_{n-1}$$

and

$$v_{n+1}^2 - v_{n+2}v_n = v_{n+1}^2 - (kv_{n+1} + v_n)v_n$$
$$= v_{n+1}(v_{n+1} - kv_n) - v_n^2 = v_{n+1}v_{n-1} - v_n^2$$

it is straightforward to prove by induction that the squares of each term of each sequence differ from the product of the adjacent terms by ± 1.

For the difference and sum of squares results, we need to show for all applicable n that

$$u_n(u_{n+1} - u_{n-1}) = u_{2n}$$
$$u_{n+1}^2 - u_n^2 = u_{2n+1}$$
$$v_n(v_{n+1} + v_{n-1}) = v_{2n}$$
$$v_{n+1}^2 + v_n^2 = v_{2n+1}.$$

These follow from an induction argument involving the relations

$$(u_{n+1}^2 - u_n^2) - ku_n(u_{n+1} - u_{n-1}) + (u_n^2 - u_{n-1}^2)$$
$$= u_{n+1}(u_{n+1} - ku_n) + u_{n-1}(ku_n - u_{n-1})$$
$$= u_{n+1}(-u_{n-1}) + u_{n-1}u_{n+1} = 0,$$
$$u_{n+1}(u_{n+2} - u_n) - k(u_{n+1}^2 - u_n^2) + u_n(u_{n+1} - u_{n-1})$$
$$= u_{n+1}(u_{n+2} - ku_{n+1}) + u_n(ku_n - u_{n-1}) = 0.$$
$$(v_{n+1}^2 + v_n^2) - kv_n(v_{n+1} + v_{n-1}) - (v_n^2 + v_{n-1}^2)$$
$$= v_{n+1}(v_{n+1} - kv_n) - v_{n-1}(kv_n + v_{n-1}) = 0,$$
$$v_{n+1}(v_{n+2} + v_n) - k(v_{n+1}^2 + v_n^2) - v_n(v_{n+1} + v_{n-1})$$
$$= v_{n+1}(v_{n+2} - kv_{n+1}) - v_n(kv_n + v_{n-1}) = 0.$$

7. The general solution of the second-order recursion $x_n = ax_{n-1} + bx_{n-2}$ is given by

$$x_n = \alpha\lambda^n + \beta\mu^n$$

in the event that the quadratic equation $t^2 = at + b$ has distinct roots λ and μ, and by

$$x_n = (\alpha n + \beta)\lambda^n$$

in the event that the quadratic equation has coincident roots λ.

Thus, in the first case,

$$x_4 x_3 x_8 = \alpha^3 \lambda^{15} + \alpha^2 \beta(\lambda^7 \mu^8 + \lambda^{11} \mu^4 + \lambda^{12} \mu^3)$$
$$+ \alpha\beta^2(\lambda^4 \mu^{11} + \lambda^3 \mu^{12} + \lambda^8 \mu^7) + \beta^3 \mu^{15}$$

while in the second

$$x_4 x_3 x_8 = (4\alpha + \beta)(3\alpha + \beta)(8\alpha + \beta)\lambda^{15}$$
$$= (96\alpha^3 + 68\alpha^2\beta + 15\alpha\beta^2 + \beta^3)\lambda^{15}.$$

Working out each of the products in this way and combining will verify the result.

8. See problem 10211 in *American Mathematical Monthly* 99 (1992), 361.

It is straightforward to check that $x_2 = (bc - ad)^2$. We establish by induction that each x_{n+1} is not only positive, but equal to $(cu_n - du_{n-1})^2$, where $u_0 = |a|$, $u_1 = |b|$ and u_r is the positive square root of x_r for $r = 1, 2, \ldots, n$. Suppose that this has been established for $n = 1, 2, \ldots, k - 1$.

Observe that

$$u_k^2 + u_{k-1}^2 d - u_{k-1}u_k c = (cu_{k-1} - du_{k-2})^2 + u_{k-1}^2 d$$
$$- u_{k-1}(cu_{k-1} - du_{k-2})c$$
$$= d(u_{k-1}^2 + u_{k-2}^2 d - u_{k-1}u_{k-2}c) = \cdots$$
$$= d^{k-1}(u_1^2 + u_0^2 d - u_1 u_0 c)$$
$$= (b^2 + a^2 d - abc)d^{k-1}.$$

Hence,

$$x_{k+1} = (c^2 - 2d)u_k^2 - d^2 u_{k-1}^2 + 2(u_k^2 + u_{k-1}^2 d - u_{k-1}u_k c)d$$
$$= (cu_k - du_{k-1})^2,$$

as desired.

9. See *Scientific American* (June, 1976), 120–125 and Henry W. Gould, *Bell and Catalan numbers: research bibliography of two special number sequences.* 6th ed., Morgantown, WV, 1985. For $n \geq 2$, the nth Catalan number is

$$\frac{1}{2n - 1}\binom{2n - 1}{n - 1} = \frac{1}{n}\binom{2n - 2}{n - 1}$$

$$= \frac{1}{n - 1}\binom{2n - 2}{n} = \frac{(2n - 2)!}{n!(n - 1)!}.$$

A straightforward but tedious calculation verifies that

$$1 + 24 \left[\frac{x_{n-1}x_{n-2}}{x_n x_{n-2} - x_{n-1}^2} \right] = (2n-1)^2.$$

A discussion of unsolved problems in which Catalan numbers arise appears in *American Mathematical Monthly* 100 (1993), 287–289.

Consider the following expressions:

$$x_1(x_1 + x_2) = x_1^2 + x_1 x_2$$

$$x_1(x_1 + x_2)(x_1 + x_2 + x_3) = x_1^3 + 2x_1^2 x_2 + x_1^2 x_3 + x_1 x_2^2 + x_1 x_2 x_3$$

$$x_1(x_1 + x_2)(x_1 + x_2 + x_3)(x_1 + x_2 + x_3 + x_4)$$
$$= x_1^4 + 3x_1^3 x_2 + 2x_1^3 x_3 + \cdots + x_1 x_2 x_3 x_4.$$

The number of terms in the expansions are respectively 2, 5, 14. In general, the number of terms (after like terms have been combined) in the expansion of $x_1(x_1 + x_2)(x_1 + x_2 + x_3) \cdots (x_1 + x_2 + \cdots + x - n)$ is the $(n+1)$th Catalan number (Problem E2972, *American Mathematical Monthly* 89 (1982), 698; 93 (1986), 217).

10. See page 355 of *Reviews in Number Theory,* 1973–1983, Volume 1A, B66–220, B66–221 (American Mathematical Society).

11. It was shown by Euler that it is impossible to have four squares in arithmetic progresssion. See Volume 2 of Dickson's *History,* page 440 (Chapter XIV) for a sketch of the argument. A method of finding three squares in arithmetic progression is given by Leonardo of Pisa (Fibonacci) as Proposition 14 in his Book of Squares (*Liber Quadratorum*), published in 1225. In the solution to problem 264 of *American Mathematical Monthly* 24 (1917), 177; 25 (1918), 123–124, a formula for three squares in arithmetic progression is given. If $y^2 - x^2 = z^2 - y^2$, then $x^2 + z^2 = 2y^2$, and so x and z have the same parity. We can write $z = v + w$ and $x = v - w$, for integers v and w, obtaining $v^2 + w^2 = y^2$. Using the general formula for pythagorean triples, we find that there are integers p and q for which $(x, y, z) = (p^2 - q^2 - 2pq, p^2 + q^2, p^2 - q^2 + 2pq)$.

In Problem E1280 in *American Mathematical Monthly* 64 (1957), 505; 65 (1958), 210–211, it is pointed out that $24+1$ and $2 \times 24+1$ are squares, as are also $2 \times 40+1$ and $3 \times 40+1$. Is it possible to find an integer x for which $x + 1$, $2x + 1$ and $3x + 1$ are all square? If so, then these numbers, along with 1, would constitute four squares in arithmetic progression. See Exercise 4(a) in Chapter 4 for the case where $x + 1$ and $3x + 1$ are square.

12. Problem E612 in *American Mathematical Monthly* 51 (1944), 162, 589–590 noted that $1110 \cdot 1111 \cdot 1112 \cdot 1113 = 1235431^2 - 1$ for any base of numeration exceeding 5, and asked for a generalization. E.P. Starke noted that this was a special case of $x(x+y)(x+2y)(x+3y) = (x^2+3xy+y^2)^2 - y^4$ with $x = b^3 + b^2 + b$ and $y = 1$.

In Problem E876 in *American Mathematical Monthly* 56 (1949), 473; 57 (1950), 186, it was pointed out that the product of four consecutive terms in arithmetic progression plus the fourth power of the difference is a perfect square and could even be a fourth power. If $x = u^2 - v^2$ and $y = 2uv + 3v^2$, then

$$x(x + y)(x + 2y)(x + 3y) = (u^2 + 3uv + v^2)^4 - y^4.$$

For example, $3 \times 10 \times 17 \times 24 = 12240 = 11^4 - 7^4$.

In a "Quickie" (Q765, *Mathematics Magazine* 63 (1990), 190, 198), E. T. Wang shows how one can use the identity

$$t(t + 1)(t + 2)(t + 3) + 1 = (t^2 + 3t + 1)^2$$

to get solutions to the diophantine equation $(x^2 - 1)(y^2 - 1) = z^2 - 1$ (or, $x^2 y^2 = x^2 + y^2 + z^2$). Simply take $(x, y, z) = (t + 1, t + 2, t^2 + 3t + 1)$.

It is not possible for the product of n consecutive nonzero integers to be a kth power of an integer for $k \geq 2$. See the articles by Paul Erdős in *Journal of the London Mathematical Society* 14 (1939), 194–198, 245–249. A discussion of this problem appears also in *American Mathematical Monthly* 47 (1940), 280–289.

A discussion of the diophantine equation

$$y^m = x(x + 1)(x + 2) \cdots (x + n - 1)$$

appears on pages 267–268 of the book, L. J. Mordell, *Diophantine equations,* Academic, 1969.

13. See the article by E. J. Barbeau in *Canadian Mathematical Bulletin* 28 (1985), 337–342.

Additional Exercises

1. In 1883, G. C. Gerono stated that if n^2 is the sum of two consecutive squares, then n is the sum of three squares of which two are consecutive (Dickson's *History*, Volume II, p. 268). Explain why this is so.

2. *Elementary number theory; easy.* Are there integers x, a, b, c, d for which $(x + 1)^2 + a^2 = (x + 2)^2 + b^2 = (x + 3)^2 + c^2 = (x + 4)^2 + d^2$?

3. Observe that $\binom{2}{2} = 1^2$ and that $\binom{6}{2} + 2\binom{6}{4} + 4\binom{6}{6} = 7^2$. Prove, generally, that

$$\sum_{i=1}^{2n-1} 2^{i-1} \binom{4n-2}{2i}$$

is a square, for every positive integer n.

Solutions

1. When I first examined this question, I came up with the following set of equations which did not appear to indicate much: $29 = 2^2 + 3^2 + 4^2$, $169 = 3^2+4^2+12^2$, $985 = 6^2+7^2+30^2$, $5741 = 11^2+12^2+74^2$ and $33461 = 86^2+87^2+136^2$. This turned out to be a wrong track. The secret is to observe that each term in the sequence $5, 29, 169, \ldots$ of numbers whose squares are equal to the sum of two consecutive squares is the sum of the squares of two consecutive terms in the sequence $1, 2, 5, 12, 29, 70, 169, 408, \ldots$ where each term is the sum of twice its predecessor plus the one before. The terms $5, 29, 169, \ldots$ occur in alternate positions. Thus we have

$$29 = 2^2 + 5^2 = 2^2 + 3^2 + 4^2 \qquad \text{since } 5^2 = 3^2 + 4^2$$

$$169 = 5^2 + 12^2 = 3^2 + 4^2 + 12^2$$

$$985 = 12^2 + 29^2 = 12^2 + 20^2 + 21^2 \qquad \text{since } 29^2 = 20^2 + 21^2$$

$$5741 = 29^2 + 70^2 = 20^2 + 21^2 + 70^2$$

and so on.

This is Problem 2751 in the *American Mathematical Monthly* 26 (1919), 73.

2. No. Analyze the situation modulo 8. (Problem E3080, *American Mathematical Monthly* 92 (1985), 215; 95 (1988), 141)

3. This is Problem 521 in *Mathematics Magazine* 36 (1963), 198; 37 (1964), 61. Observe that the sum is equal to

$$\frac{1}{4}\left[\left(1+\sqrt{2}\right)^{4n-2} + \left(1-\sqrt{2}\right)^{4n-2} - 2\right]$$

$$= \left[\frac{(1+\sqrt{2})^{2n-1} + (1-\sqrt{2})^{2n-1}}{2}\right]^2.$$

It is readily seen that the quantity in square brackets is an integer, since the terms in $\sqrt{2}$ cancel out. Let us examine it in more detail. Let $u = 1 + \sqrt{2}$ and $v = 1 - \sqrt{2}$. Then $u^2 + v^2 = 6$ and $u^2v^2 = 1$, so that u^2 and v^2 are roots of the quadratic equation $t^2 = 6t - 1$. From this, we can verify that $u^{2n-1} = 6u^{2n-3} - u^{2n-5}$ and $v^{2n-1} = 6v^{2n-3} - v^{2n-5}$. Thus, we find that

$$w_n = \frac{u^{2n-1} + v^{2n-1}}{2}$$

satisfies the recursion

$$w_1 = 1 \quad w_2 = 7 \quad w_n = 6w_{n-1} - w_{n-2} \ (n \geq 3).$$

Thus, the sum of the problem is equal to w_n^2, where $\{w_n\}$ is the sequence $\{1, 7, 41, 239, 1393, \ldots\}$.

CHAPTER **4**

Pell's Equation

A quadratic diophantine equation

It is not possible for one square to be twice another. This is a fact known to the ancient Greeks. However, there are many squares which differ by only one from twice another square:

$$3^2 = 2 \times 2^2 + 1; \quad 7^2 = 2 \times 5^2 - 1; \quad 17^2 = 2 \times 12^2 + 1; \ldots.$$

You may recognize the numbers in these examples as arising from sequences A and B in Chapter 3 (p. 60). In fact, it turns out that *all* the solutions in positive integers x and y of the equations $x^2 - 2y^2 = 1$ and $x^2 - 2y^2 = -1$ can be found by taking x and y to be corresponding terms of the two sequences.

Similarly, it is not possible for one square to be three times another. However, there are many squares which differ from three times another by only 1, such as 7^2, which is 1 more than 3×4^2. Thus, $(x, y) = (7, 4)$ is a solution of the equation $x^2 - 3y^2 = 1$.

However, any square of an even number is four times the square of another number, namely half the value of the even number.

In general, it is possible for one square to be d times another exactly when d is itself a perfect square. If d is a positive integer, but not a square, then $x^2 = dy^2$ can never be solved for integers x and y. However, in this situation, the equation

$$x^2 - dy^2 = 1$$

has an infinite number of integer solutions (x, y). This equation is known as *Pell's equation*.

Sometimes, it is easy to find a solution by trial and error. For example, $x^2 - 60y^2 = 1$ is satisfied by $x = 31$, $y = 4$. Other times, it is practically impossible to find a solution by trial and error. Try the equation $x^2 - 61y^2 = 1$.

Fortunately, there is a simple algorithm to find a solution which is amenable to a calculator.

Here it is:

1. Let u be the square root of d.
2. Suppose $u = v + w$, where v is a positive integer and $0 < w < 1$. Record v and subtract v from u to get the fractional part w of u.
3. Take the reciprocal of w. Call this reciprocal u and repeat step (2) with this new value of u.

For $d = 61$, we find that $\sqrt{61} = 7.81024967591$. We record 7 and reciprocate 0.81024967591 to get $1/0.81024967590 = 1.2341874730$.

While the algorithm is very easy to carry out with a pocket calculator, its recursive nature will permit any rounding or truncation errors to worsen to the point where even the term before the decimal point is affected. This is illustrated in the following table for $d = 61$, in which at each stage we have indicated the ever-widening bounds in which the true values lie. For example, if we accept 7.81024967591 as the first entry, then the second entry is a little greater than 1.23418747299, while if we accept 7.81024967590 as the first entry, the second is a little less than 1.23418747301. The true value is somewhere in between.

Upper bound	Lower bound
7.81024967591	7.81024967590
1.23418747301	1.23418747299
4.27008322535	4.27008322497
3.70256242354	3.70256241832
1.42336107644	1.42336106585
2.36204998680	2.36204992771
2.76204999218	2.76204954138
1.31225064212	1.31224986583
3.20256342575	3.20255546380
4.93691940588	4.93672535550
1.06754876884	1.06732766310
14.852735918	14.8401188192
1.24359730941	1.17269599988
5.79052207749	4.10513565367

At this point, we stop, since the integer part of the number is no longer known with certainty. However, we have enough to do the job.

Form the following table. The first column consists of the integer parts of the numbers in the foregoing table insofar as we know them for sure. After the number 14 (double the original 7), there is a theorem which tells us that the next block of entries repeat those between 7 and 14. The top entry in the

x column agrees with the first number in the left column. The second x entry is one more than the product of the first two numbers in the left column. The top entry in the y column is 1. The second y entry is the same as the second number in the left column. From the third entry on, there is a common rule of formation for both the x and y columns:

multiply the corresponding number in the left column by the previous entry in the column and add the entry before

For example, the entry 24079 some way down in the x column is equal to 4×5639 plus 1523. Similarly, the y-entry 3083 equals 4×722 plus 195.

number	x	y	$x^2 - 61y^2$
7	7	1	-12
1	8	1	3
4	39	5	-4
3	125	16	9
1	164	21	-4
2	453	58	5
2	1070	137	-9
1	1523	195	4
3	5639	722	-3
4	24079	3083	12
1	29718	3805	-1
14	440131	56353	12
1	469849	60158	-3
4	2319527	296985	4
3	7428430	951113	-9
1	9747957	1248098	5
2	26924344	3447309	-5
2	63596645	8142716	9
1	90520989	11590025	-4
3	335159612	42912791	3
4	1431159437	183241189	-12
1	1766319049	226153980	1

We are finished our search. The smallest values of x and y for which $x^2 - 61y^2 = 1$ are 1766319049 and 226153980.

The process of getting the integer parts can be carried out more tediously, but exactly, without a calculator, using the algebra of surds. For $d = 61$, we have

$$\sqrt{61} = 7 + (\sqrt{61} - 7) = 7 + \frac{12}{\sqrt{61} + 7}$$

$$\frac{\sqrt{61}+7}{12} = 1 + \frac{\sqrt{61}-5}{12} = 1 + \frac{3}{\sqrt{61}+5}$$

$$\frac{\sqrt{61}+5}{3} = 4 + \frac{\sqrt{61}-7}{3} = 4 + \frac{4}{\sqrt{61}+7}$$

$$\frac{\sqrt{61}+7}{4} = 3 + \frac{\sqrt{61}-5}{4} = 3 + \frac{9}{\sqrt{61}+5}$$

and so on. The reader may wish to continue the table and generate the sequence $\{7, 1, 4, 3, 1, 2, 2, 1, \ldots\}$ of integer parts. Note that when the surds $(\sqrt{61} - a)/b$ are reciprocated, it turns out that b is always a divisor of $61 - a^2 = (\sqrt{61} - a)(\sqrt{61} + a)$; there are some interesting connections among the integers occurring as numerators and denominators of the surds.

There is a different strategy for generating rational approximations to $\sqrt{d}$ and solutions to $x^2 - dy^2 = 1$. It is based on the idea that if a/b and c/d are two positive vulgar fractions, then the fraction $(a + c)/(b + d)$ lies between them in value. Thus, if a/b is an underapproximation of $\sqrt{d}$ (so that $a^2 - db^2 < 0$) and c/d is an overapproximation (so that $a^2 - db^2 > 0$), then $(a + c)/(b + d)$ should be a better approximation than at least one of a/b and c/d.

Applying this to $d = 13$ and starting with the underapproximation $3 = 3/1$ and the overapproximation $4 = 4/1$ for the square root of 13, we are led to the next approximation $7/2 = (3 + 4)/(1 + 1)$. Since $7^2 - 13 \times 2^2 = -3 < 0$, $7/2$ is an underapproximation, so we find the next approximation $11/3$ between $7/2$ and $4/1$. Continue on.

We can present the process in the form of a table:

x	y	$x^2 - 13y^2$
3	1	-4
4	1	3
7	2	-3
11	3	4
18	5	-1
29	8	9
47	13	12
65	18	13
83	23	12
101	28	9
119	33	4
137	38	-3
256	71	3
393	109	-4
649	180	1

In each instance from the third row on, the next row can be found by taking the x and y values of the previous row and adding them to the x and y values of the last preceding row for which $x^2 - dy^2$ has opposite sign. Eventually, you will reach x and y for which $x^2 - dy^2 = 1$. In the example, we find that $649^2 - 13 \times 180^2 = 1$.

For $d = 61$, the table begins:

x	y	$x^2 - 61y^2$
7	1	-12
8	1	3
15	2	-19
23	3	-20
31	4	-15
39	5	-4
47	6	13
86	11	15
125	16	9
164	21	-5

The reader may wish to continue this table and compare it with entries in the tables produced by the other method. In particular, note the number of consecutive values of $x^2 - 61y^2$ in the third column that maintain the same sign. See the Appendix (p. 165) to this chapter for further discussion.

There is a pictorial representation of the last process that depends on the algebraic identity

$$\left[(a+u)^2 - (b+v)^2d\right] + \left[(a-u)^2 - (b-v)^2d\right]$$
$$= 2\left[(a^2 - b^2d) + (u^2 - v^2d)\right] \qquad (*)$$

Suppose that $m = a^2 - b^2d$ is the last value we have so far in the third column. Let h be the most recent value of the same sign and $p = u^2 - v^2d$ the most recent value of the opposite sign. In the foregoing table for $d = 61$, for example, $m = -5$, $h = -4$, and $p = 9$. We have $h = (a-u)^2 - (b-v)^2d$. The next entry of the table is $(a+u)^2 - (b+v)^2d$, which by $(*)$ is equal to $2(m+p) - h$.

We can think of the values of $x^2 - dy^2$ as sitting in cells surrounded by edges, where the values along a given edge are related to the values at the end of the edge by Figure 4.1.

The sum of the two numbers at the end of the dividing edge is twice the sum of the two numbers on either side. Thus, knowing h, m, p, we can fill in the cell $2(m+h) - p$. Beginning with the first three values in column 3, we can compute the rest of the values along the cells. For the case $d = 61$,

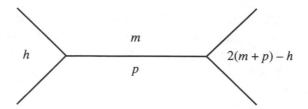

FIGURE 4.1

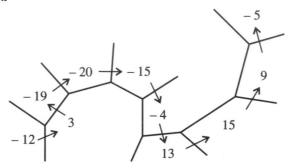

FIGURE 4.2

we start with $m = -19$, $h = -12$, $p = 3$ and get Figure 4.2, with the arrows indicating the order of computation.

The arrows pass to cells adjacent to the polygon of edges that separate positive and negative values, passing occasionally from one side to the other.

The reader may wish to try to generate solutions to $x^2 - dy^2 = 1$ for other positive integer values of d. You can rest assured that in most cases you will get to the answer much quicker than for $d = 61$.

Applications of Pell's equation

Pell's equation turns up in the investigation of a great many numerical problems. Consider, for example, the running totals of the natural numbers:

$$1, \quad 3 = 1 + 2, \quad 6 = 1 + 2 + 3, \quad 10 = 1 + 2 + 3 + 4, \ldots.$$

If we continue this further, we note that sometimes the number is a perfect square. For example,

$$1 + 2 + 3 + 4 + 5 + 6 + 7 + 8 = 36 = 6^2.$$

Is this an unusual occurrence, or can we find some sort of rule for this to happen? **(1)**

Note that, for a general positive integer n,

$$1 + 2 + 3 + \cdots + n$$
$$= \frac{1}{2}(1 + 2 + \cdots + n - 1 + n) + (n + n - 1 + \cdots + 2 + 1)$$
$$= \frac{1}{2}((1 + n) + (2 + n - 1) + \cdots + (n - 1 + 2) + (n + 1))$$
$$= \frac{n(n+1)}{2}.$$

The sum of the first n natural numbers is a square m^2 precisely when $(1/2)n(n + 1) = m^2$, or equivalently,

$$(2n + 1)^2 - 2(2m)^2 = 1.$$

With $x = 2n + 1$ and $y = 2m$, this is the pellian equation $x^2 - 2y^2 = 1$. Solving this, we find values for (m, n), including $(49, 35)$:

$$1 + 2 + \cdots + 48 + 49 = 35^2.$$

A related question is whether it is possible that, for positive integers, m and n,

$$(1 + 2 + \cdots + m)^2 = 1 + 2 + \cdots + n.$$

With $u = 2n + 1$ and $v = 2m + 1$, this leads to the equation

$$u^2 - 2\left(\frac{v^2 - 1}{4}\right)^2 = 1,$$

which, despite its Pell-like appearance, is actually an equation of the fourth degree. There are only two solutions, yielding the possibilities $(m, n) = (1, 1)$ and $(3, 8)$ **(2)**.

In the Exercises at the end of this chapter, other problems are listed. We will mention another problem that involves cubes and relates to the material in Chapter 2. Consider integers n which can be expressed as the sum of four (not necessarily positive) integer cubes: $x^3 + y^3 + z^3 + w^3 = n$.

An example is 2, which can be so expressed in several ways:

$$2 = (-5)^3 + (-1)^3 + 4^3 + 4^3 = 0^3 + 0^3 + 1^3 + 1^3$$
$$= (-6)^3 + (-5)^3 + 0^3 + 7^3 = (-8)^3 + (-6)^3 + 1^3 + 9^3.$$

Since any equation of the form $u^3 + v^3 + w^3 = r^3$ can be rewritten as $(-r)^3 + u^3 + v^3 + w^3 = 0$, we see from Chapter 2 that 0 can be written as the sum of four cubes in infinitely many ways.

In fact, there is a general criterion that guarantees infinitely many representations **(3)**:

let n be an integer; then the equation

$$x^3 + y^3 + z^3 + w^3 = n$$

has infinitely many solutions (x, y, z, w) if there can be found one solution $(x, y, z, w) = (a, b, c, d)$ with $(a + b)(c + d)$ negative and with either $a \neq b$ or $c \neq d$.

When $n = 2$, the solution $(x, y, z, w) = (-5, -1, 4, 4)$ fulfils the requirements. In the discussion of this result in the Notes, we will show how to generate infinitely many solutions for $n = 2$.

Pell's equation and second-order recursions

Pell's equation can be related to second-order recursions. For example, consider the following sequences:

n	y_n	x_n
0	0	1
1	1	3
2	6	17
3	35	99
4	204	577
5	1189	3363

Both the x and y sequences have the same law of formation: from the third term on, each term is 6 times the previous term minus the one before. The importance of this pairing lies in the fact that, for each n, $x_n^2 - 8y_n^2 = 1$.

We can get other similar pairings. Let k be any positive integer, and let $y_0 = 0$, $y_1 = 1$ and, for $n \geq 2$, $y_n = ky_{n-1} - y_{n-2}$. Now define $x_0 = 2$ and, for $n \geq 1$, $x_n = ky_n - 2y_{n-1}$. Then we find that $x_n^2 - (k^2 - 4)y_n^2 = 4$ **(4)**. When k is even, then so are all the x_n, and we can derive solutions to equations of the form $x^2 - dy^2 = 1$ by dividing through the equation from the sequences by 4.

Another similar pairing occurs by taking, for any positive integer k, $y_0 = 0$, $y_1 = 1$ and, for $n \geq 2$, $y_n = ky_{n-1} + y_{n-2}$. If $x_0 = 2$ and, for $n \geq 1$, $x_n = ky_n + 2y_{n-1}$, we have that $x_n^2 - (k^2 + 4)y_n^2 = (-1)^n \times 4$.

Other pairings of sequences give solutions to other types of Pell's equation with the number 1 on the right side replaced by other integers. For example, in the following table, both sequences continue indefinitely in both directions and each term is three times its predecessor minus the one before.

We always have $x_n^2 - 5y_n^2 = 44$:

y_n	x_n
-19	43
-7	17
-2	8
1	7
5	13
14	32
37	83

Pell's equations of higher degree

Note that $x^2 - dy^2 = (x + y\sqrt{d})(x - y\sqrt{d})$ and that $\sqrt{d}$ and $-\sqrt{d}$ are the two roots of the quadratic equation $t^2 - d = 0$. Now let d be a positive integer that is not a perfect cube. The three roots of the cubic equation $t^3 - d = 0$ are $d^{1/3}$, $\omega d^{1/3}$ and $\omega^2 d^{1/3}$, where $d^{1/3}$ is the positive cube root of d and ω is an "imaginary" cube root of 1, i.e., $\omega^3 = 1$ and $\omega^2 + \omega + 1 = 0$. We form the product of the terms $(x + \theta y + \theta^2 z)$ where θ assumes the values of the three roots of $t^3 - d = 0$:

$$(x + d^{1/3}y + d^{2/3}z)(x + \omega d^{1/3}y + \omega^2 d^{2/3}z)(x + \omega^2 d^{1/3}y + \omega d^{2/3}z)$$

$$= x^3 + dy^3 + d^2 z^3 - 3dxyz$$

$$= \frac{1}{2}(x + d^{1/3}y + d^{2/3}z) \times$$

$$\left[(x - d^{1/3}y)^2 + d^{2/3}(y - d^{1/3}z)^2 + (x - d^{2/3}z)^2\right].$$

The cubic analogue of Pell's equation is

$$x^3 + dy^3 + d^2 z^3 - 3dxyz = 1.$$

If (x, y, z) is a solution in large positive integers, then from the factorization of the left side, $x + d^{1/3}y + d^{2/3}z$ would be large and each of $x - d^{1/3}y$, $y - d^{1/3}z$ and $x - d^{2/3}z$ would have to be very small. Thus we look for x/y and y/z to be close to the cube root of d.

Keeping this idea in mind, we try to adapt the table for the quadratic method to obtaining a solution for the cubic case. We form a table with four initial rows that correspond to the four possible sign pairs for $(x^3 - dy^3, y^3 - dz^3)$. In each row $z = 1$. In the first row, we select the largest possible y for which $y^3 - dz^3$ is negative, and then the largest possible x for which $x^3 - dy^3$ is negative. For the second row, we select y as before, but take the smallest possible x for which $x^3 - dy^3$ is positive. In the third and fourth rows, choose the smallest y for which $y^3 - dz^3$ is positive. In the third row,

x is largest for which $x^3 - dy^3$ is negative and in the fourth row, smallest for which $x^3 - dy^3$ is positive.

Once we have the initial four rows, the vector (x, y, z) in each subsequent row is found by adding the vector of values for the previous row to the vector of the last of the earlier rows for which both $x^3 - dy^3$ and $y^3 - dz^3$ differ in sign. For the case $d = 2$, the table starts like this:

(x, y, z)	$x^3 - 2y^3$	$y^3 - 2z^3$	$x^3 + 2y^3 + 4z^3 - 6xyz$
$(1, 1, 1)^*$	$-$	$-$	1
$(2, 1, 1)$	$+$	$-$	2
$(2, 2, 1)$	$-$	$+$	4
$(3, 2, 1)$	$+$	$+$	11
$(4, 3, 2)$	$+$	$+$	6
$(5, 4, 3)^*$	$-$	$+$	1
$(7, 5, 4)$	$+$	$-$	9
$(12, 9, 7)$	$+$	$+$	22
$(13, 10, 8)$	$+$	$-$	5
$(18, 14, 11)$	$+$	$+$	12
$(19, 15, 12)^*$	$+$	$-$	1

The asterisks indicate solutions to the cubic Pell's equation

$$x^3 + 2y^3 + 4z^3 - 6xyz = 1.$$

The reader is invited to continue the table to see what other solutions are picked up.

Unlike the quadratic case, the process for the cubic does not seem to have nice structural properties. While it might not work for every value of d, it seems to generate a solution surprisingly often. When it does generate a solution, it does not pick up every solution, but actually misses quite a few. Any reader with access to highpowered computer software like *Mathematica* is earnestly invited to experiment. See the Appendix for further details.

Pell-like equations of higher degree

A natural generalization of the quadratic Pell's equation is $x^n - dy^n = k$, where $n \geq 3$, especially when $k = \pm 1$. Solutions exist only for certain values of n and d. When $n = 3$, we have, for instance,

$$10^3 - 37 \times 3^3 = 1 \quad 7^3 - 43 \times 2^3 = -1 \quad 9^3 - 91 \times 2^3 = 1.$$

For $n = 4$, examples can be derived from the ordinary quadratic case. Suppose that $u^2 - cv^2 = 1$, where c, u, v are integers and $2c$ is a multiple of

v^2. Let $d = c^2 + (2c/v^2)$. Then

$$u^4 - dv^4 = (1 + cv^2)^2 - \left(c^2 + \frac{2c}{v^2}\right)v^4 = 1.$$

When $v = 1$, we can take $c = u^2 - 1$, whence $d = u^4 - 1$. Then $(x, y) = (u, 1)$ satisfies both the equations $x^2 - (u^2 - 1)y^2 = 1$ and $x^4 - (u^4 - 1)y^4 = 1$. More interesting are the cases with larger values of v. When $v = 2$, we can relate c and u by $4c = u^2 - 1$. When $u = 2w + 1$ is an odd number, then $c = w(w + 1)$ is an even integer and

$$d = c^2 + \frac{c}{2} = \left[w(w + 1)\right]^2 + \frac{1}{2}w(w + 1).$$

Thus, $(x, y) = (2w + 1, 2)$ satisfies both

$$x^2 - w(w + 1)y^2 = 1$$

and

$$x^4 - \frac{w(w + 1)}{2}\left[w^2 + (w + 1)^2\right]y^4 = 1.$$

In particular,

$$3^4 - 5 \times 2^4 = 1, \quad 5^4 - 39 \times 2^4 = 1,$$

$$7^4 - 150 \times 2^4 = 1, \quad 9^4 - 410 \times 2^4 = 1.$$

When $v = 3$, take $9c = u^2 - 1$. When $u^2 - 1$ is divisible by 81, then c and d are integers. Thus, taking u to be 80 and 82 yields the examples

$$80^4 - 505679 \times 3^4 = 1 \quad \text{and} \quad 82^4 - 558175 \times 3^4 = 1.$$

Examples with $v = 4$ are

$$63^4 - 61535 \times 4^4 = 1 \quad \text{and} \quad 65^4 - 69729 \times 4^4 = 1.$$

In general, for given d, $x^4 - dy^4 = \pm 1$ has at most one solution in positive integers. Even if we relax the condition on the power of x, we find that for a wide class of values of d, $x^2 - dy^4 = -1$ has at most two positive integer solutions. For example, $x^2 - 2y^4 = -1$ is satisfied only by $(x, y) = (1, 1)$ and $(239, 13)$ **(5)**.

One method of constructing solutions to new mixed power equations from old ones is described by A. M. S. Ramasamy **(6)**. Suppose that m is a positive integer and that

$$u^2 - 2v^m = 1.$$

Then, if $x = u(v^m - 1)$, $y = v$ and $d = u^2 - 4$, then

$$x^2 - dy^{2m} = u^2 v^{2m} - 2u^2 v^m + u^2 - u^2 v^{2m} + 4v^{2m}$$
$$= u^2 - 2v^m(u^2 - 2v^m) = u^2 - 2v^m = 1.$$

Similarly, if $u^2 - 2v^m = -1$, then $(d; x, y) = (u^2 + 4; u(v^m + 1), v)$ satisfies $x^2 - dy^{2m} = -1$. For example, from $3^2 - 2 \times 2^2 = 1$, we obtain $9^2 - 5 \times 2^4 = 1$; from $7^2 - 2 \times 5^2 = -1$, we obtain $182^2 - 53 \times 5^4 = -1$ and from $239^2 - 2 \times 13^4 = -1$, we obtain $6826318^2 - 57125 \times 13^8 = -1$.

When dealing with mixed powers, we can take $d = 1$ and ask about solutions of $x^m - y^n = k$, where $m, n \geq 2$ and k is small. For example, $3^2 - 2^3 = 1$, $3^3 - 5^2 = 2$, $2^7 - 5^3 = 3$, $5^3 - 11^2 = 4$. Can every integer be realized as the difference of two distinct powers above the first? It was conjectured by Catalan in 1842 that $x^m - y^n = 1$ holds only for $(x, y; m, n) = (3, 2; 2, 3)$, where all the variables are integers exceeding 1. It had been found in the previous century by Euler (1707–1783) that $x^2 - y^3 = 1$ has a single solution; since then other specific conditions on m and n guaranteeing an absence of solutions (x, y) have been found **(7)**.

Notes

The theory of Pell's equation and approximation of nonrational numbers by rationals using continued fractions is standard fare in a university number theory course. Suppose that $(x, y) = (u, v)$ is the smallest solution of the pellian equation $x^2 - dy^2 = 1$ in positive integers. Then the entire set of solutions in integers is given by $(x, y) = (u_n, v_n)$ where

$$(u + v\sqrt{d})^n = u_n + v_n\sqrt{d}$$

and n ranges over the set of positive and negative integers. The pair (u, v) is called the *fundamental solution*.

To solve the equation $x^2 - dy^2 = k$, determine a particular solution $(x, y) = (p, q)$; other solutions can be constructed by combining such a solution with any solution of $x^2 - dy^2 = 1$. Thus, if (x_n, y_n) is determined by

$$x_n + y_n\sqrt{d} = (p + q\sqrt{d})(u_n + v_n\sqrt{d})$$

where u_n and v_n are given as above, then $x_n^2 - dy_n^2 = k$.

We thus have, for each integer n,

$$x_{n+1} + y_{n+1}\sqrt{d} = (u + v\sqrt{d})(x_n + y_n\sqrt{d}).$$

Expanding out, and collecting rational and nonrational terms, we find that this is equivalent to the pair of equations:

$$x_{n+1} = ux_n + dvy_n$$

$$y_{n+1} = vx_n + uy_n.$$

In vector form, this becomes $X_{n+1} = MX_n$, where X_n is the column vector with entries x_n and y_n and M is the 2×2 matrix whose first row consists of the numbers u and dv and whose second row consists of v and u. The determinant of this matrix is equal to 1 and its characteristic equation is $t^2 - 2ut + 1 = 0$. By the Cayley–Hamilton Theorem, $M^2 - 2uM + I = 0$. It follows that, for each positive integer n,

$$X_{n+2} - 2uX_{n+1} + X_n = (M^2 - 2uM + I)X_n = O \cdot X_n = O.$$

From this, we find that

$$x_{n+2} = 2ux_{n+1} - x_n$$

$$y_{n+2} = 2uy_{n+1} - y_n.$$

For further information, consult

Z. I. Borevich and I. R. Shafarevich, *Number theory*, Academic, 1966. Chapter 2.

G. H. Hardy and E. M. Wright, *An introduction to the theory of numbers*, Oxford. Chapter X.

W. J. LeVeque, *Topics in number theory*, Volume I Addison-Wesley, 1956. Chapters 8 and 9.

I. Niven and H. S. Zuckerman, *An introduction to the theory of numbers*, Wiley, 1960, 1966, 1972, Chapter 7.

H. Rademacher, *Higher mathematics from an elementary point of view*, Birkhäuser, 1983, Chapter 7.

Good surveys of results are

L. J. Mordell, *Diophantine equations*, Academic, 1969.

André Weil, *Number theory; an approach through history from Hammurapi to Legendre*, Birkhäuser, 1983.

Reviews in Number Theory, American Mathematical Society. Two sets have already appeared, covering the periods 1940–1972 and 1973–1983. Material pertinent to this chapter appears in Volume 2 of each set.

1. This is posed as a "Quickie" (Q633) in *Mathematics Magazine* 49 (1976), 96, 101.

2. See page 268 of Mordell's book cited above. For an expository discussion of techniques for solving diophantine equations, see the article by R. J.

Stroeker in *American Mathematical Monthly* 91 (1984), 385–392, where particular attention is paid to the modified pellian equation $(x^2 - 2)^2 - 2y^2 = 2$.

3. See page 58 of Mordell's book. To demonstrate how the result works for the case $n = 2$, let

$$x = -5 + s, \quad y = -1 - s, \quad z = 4 + t, \quad w = 4 - t.$$

Plugging this into $x^3 + y^3 + z^3 + w^3 = 2$ yields

$$-18s^2 + 72s + 24t^2 = 0$$

which simplifies to

$$(2t)^2 - 3(s - 2)^2 = -12.$$

Set $u = 2t$, $v = s - 2$. Thus, we need to solve the equation $u^2 - 3v^2 = -12$ for even values of u; an obvious solution is $(u, v) = (0, 2)$.

Infinitely many solutions $(u, v) = (u_k, v_k)$ are given by

$$u_k + v_k \sqrt{3} = (2\sqrt{3})(2 + \sqrt{3})^k.$$

We have $(u_0, v_0) = (0, 2)$ and

$$u_{k+1} + v_{k+1}\sqrt{3} = (u_k + v_k\sqrt{3})(2 + \sqrt{3})$$

so that

$$u_{k+1} = 2u_k + 3v_k \quad \text{and} \quad v_{k+1} = u_k + 2v_k.$$

Here are a few values of (x, y, z, w) for which $x^3 + y^3 + z^3 + w^3 = 2$:

(u, v)	(s, t)	(x, y, z, w)
$(0, 2)$	$(4, 0)$	$(-1, -5, 4, 4)$
$(6, 4)$	$(6, 3)$	$(1, -7, 7, 1)$
$(24, 14)$	$(16, 12)$	$(11, -17, 16, -8)$
$(90, 52)$	$(54, 45)$	$(49, -55, 49, -41)$
$(336, 194)$	$(196, 168)$	$(191, -197, 172, -164)$

See also the article by Klarner in *American Mathematical Monthly* 74 (1967), 531–537.

4. The following set of exercises will indicate the general setting for the phenomenon described.

Suppose that t_0 and t_1 are given, and that $t_n = at_{n-1} + bt_{n-2}$ for $n \geq 0$ and that $\delta = a^2 + 4b$ (the discriminant of the characteristic polynomial

$t^2 - at - b$ for the recursion). Let

$$r_n = at_n + 2bt_{n-1} = t_{n+1} + bt_{n-1}.$$

(a) Prove that

$$t_n^2 - t_{n+1}t_{n-1} = -b(t_{n-1}^2 - t_n t_{n-2}) = (-1)^{n-1}b^{n-1}(t_1^2 - t_2 t_0)$$

and deduce that

$$r_n^2 - \delta t_n^2 = (-1)^n[t_1^2 - t_2 t_0]4b^n.$$

(b) Consider the special case: $b = -1$, $t_0 = 0$, $t_1 = 1$, $a = 2c + 1$ (an odd integer). Establish that t_n and r_n are each even if and only if n is a multiple of 3. Show that, for each such integer n, $(x, y) = (\frac{1}{2}r_n, \frac{1}{2}t_n)$ satisfies the pellian equation $x^2 - (a^2 - 4)y^2 = 1$.

(c) Consider the special case: $b = -1$, $t_0 = 0$, $t_1 = 1$, $a = 2c$ (an even number). Verify that $(x, y) = (ct_n - t_{n-1}, t_n)$ satisfies the pellian equation $x^2 - (c^2 - 1)y^2 = 1$.

(d) Carry out an analysis similar to that in (b) and (c) for the case that $b = 1$, $t_0 = 0$, $t_1 = 1$.

(e) Let d be a positive nonsquare integer, k be an integer, $u^2 - dv^2 = 1$, $x_0^2 - dy_0^2 = k$. Then, let $x_n + y_n\sqrt{d} = (x_0 + y_0\sqrt{d})(u + v\sqrt{d})^n$, so that $x_{n+1} = 2ux_n - x_{n-1}$ and $y_{n+1} = 2uy_n - y_{n-1}$, as noted above. Use $y_{n+1} = vx_n + uy_n$ to deduce that $2vx_n = 2uy_n - 2y_{n-1}$. Observe that we can put this in the framework of the earlier part of this problem by taking $a = 2u$, $b = -1$, $t_n = y_n$, $r_n = 2vx_n$ and $\delta = 4dv^2 = a^2 - 4$.

5. See Chapter 28 of Mordell's book.

6. *Indian Journal of Pure and Applied Mathematics* 25 (1994), 577–581.

7. Mordell discusses Catalan's conjecture on pages 300–302 of his book. R. Tijdeman reviews the history of the conjecture and establishes that there are only finitely many possible solutions in $(x, y : m, n)$ in *Acta Arithmetica* 29 (1976), 197–209. A recent article by P. Ribenboim in *American Mathematical Monthly* 103 (1996) 529–538 reviews progress on Catalan's conjecture and other number theoretic problems. In Problem 2927 in *American Mathematical Monthly* 28 (1921), 393; 30 (1923), 81, the reader is asked to show that $3^x - 2^y = \pm 1$ has the sole solution $(x, y) = (2, 3)$. For $3^x - 2^y = -1$, observe that y must be even. Setting $y = 2z$ and $2^z = 3m \pm 1$, obtain $3^x = (3m \pm 1)^2 - 1 = 3m(3m \pm 2)$. For $3^x - 2^y = 1$ and $y \geq 2$, note that x is even. Setting $x = 2w$ and $3^w = 2n+1$, obtain $2^y = (2n+1)^2 - 1 = 4n(n+1)$. The desired conclusion follows easily.

In Problem E1221, *American Mathematical Monthly* 63 (1956), 420; 64 (1957), 110, it is required to show that $2^n - x^m = 1$ is impossible for positive integers n, x, m unless $x = n = 1$. Let $n \geq 2$. Since $2^n - 1 \equiv 3$ (mod 4), $2^n - 1$ cannot be square. Hence m is odd. But then the second factor of

$$2^n = x^m + 1 = (x+1)(x^{m-1} + x^{m-2} + \cdots + x + 1)$$

is odd and we get a contradiction.

Exercises

The equation $x^2 - dy^2 = k$ arises in many simple problems about powers of numbers. In Chapter 1, Note **(13)**, the problem of finding sums of consecutive odd cubes equal to squares led to the equation $m^2 - 2n^2 = -1$. Chapter 2, Note **(16)**, dealt with the determination of values of n for which $(n+1)(7n+1)$ was square; this led to the equation $r^2 - 7m^2 = 9$. One way of obtaining solutions to the diophantine equation given in Exercise 10(b) of Chapter 2 led to a pellian equation. In this chapter, further examples have been given.

In the following exercises, show how the question entails the solving of an equation of the form $x^2 - dy^2 = k$ and hence obtain an answer.

1. My house is on a street where the numbers $1, 2, 3, \ldots$ run consecutively with no gaps. My own number has three digits. Curiously, the sum of all the house numbers less than mine is the same as the sum of all numbers greater than mine. What is my house number and how many houses are there on my street?

2. There are n marbles in a jar and r of them are red. Two marbles are drawn at random, without replacement. The probability that both are red is $1/2$. What are the possible values of (n, r)?

3. A number is triangular if and only if it is of the form

$$1 + 2 + 3 + \cdots + n = \frac{1}{2}n(n+1).$$

(a) Show that the sum of two consecutive triangular numbers is always square.

(b) Find three sets of three consecutive triangular numbers that sum to a perfect square.

(c) Determine a base b for which the number $(111 \cdots 1)_b$ with k digits all equal to 1 in base b numeration is triangular, regardless of the positive integer k.

4. (a) Show that there are infinitely many positive integers a such that $a + 1$ and $3a + 1$ are perfect squares.

(b) Let $\{a_n\}$ be the increasing sequence of all solutions of (a). Show that $a_n a_{n+1} + 1$ is also a perfect square for each n.

5. (a) Show that two consecutive integers can be written in the form $(x - 1)/2$ and $(x + 1)/2$, where x is an odd integer.

(b) Prove that $((x - 1)/2, (x + 1)/2, y)$ is a pythagorean triple if and only if $x^2 - 2y^2 = -1$.

(c) Use (b) to obtain infinitely many pythagorean triples whose smallest terms are consecutive.

6. Show that there are infinitely many sets of three consecutive integers such that the sum of the square of the first, twice the square of the second and three times the square of the third is a square.

7. Determine those positive integers n for which $1^5 + 2^5 + \cdots + n^5$ is a perfect square.

8. A number is said to be *powerful* if the square of any of its prime divisors also divides it. Thus, no prime divides a powerful number to the first power only. Two pairs of consecutive powerful numbers are $(288, 289) = (2^5 \cdot 3^2, 17^2)$ and $(675, 676) = (3^3 \cdot 5^2, 2^2 \cdot 13^2)$.

(a) By showing that $x^2 - 5y^2 = -1$ has infinitely many solutions in positive integers for which y is a multiple of 5, deduce that there are infinitely many pairs of consecutive powerful numbers.

(b) Determine a pair of consecutive powerful numbers, neither of which is a square.

Solutions

1. Suppose that my house number is m and there are n houses on the street. Then

$$1 + 2 + \cdots + (m - 1) = (m + 1) + \cdots + n.$$

This reduces to $2m^2 = n^2 + n$ or $(2n + 1)^2 - 2(2m)^2 = 1$, an equation of the form $x^2 - 2y^2 = 1$ where $x = 2n + 1$ and $y = 2m$. Solutions (x, y) are $(3, 2)$, $(17, 12)$, $(99, 70)$, $(577, 408)$, $(3363, 2378)$, ... The pertinent solution is $2n + 1 = 577$, $2m = 408$, from which we conclude that my house is number 204 on a street of 288 houses. See also note **(7)** in Chapter 2.

2. There are $n(n - 1)$ ways of drawing two marbles from the jar and $r(r - 1)$ ways of both being red. Hence

$$\frac{r(r - 1)}{n(n - 1)} = \frac{1}{2}.$$

This leads to $(2n-1)^2 - 2(2r-1)^2 = -1$, which has the form $x^2 - 2y^2 = -1$ with $x = 2n - 1$, $y = 2r - 1$. Solutions $(x, y; n, r)$ are $(7, 5; 4, 3)$, $(41, 29; 21, 15)$, $(239, 169; 120, 85)$, ...

3. (a) The sum of two consecutive triangular numbers is of the form

$$\frac{1}{2}n(n-1) + \frac{1}{2}n(n+1) = n^2.$$

(b) The sum of three consecutive triangular numbers is of the form

$$\frac{1}{2}n(n-1) + \frac{1}{2}n(n+1) + \frac{1}{2}(n+1)(n+2) = \frac{1}{2}(3n^2 + 3n + 2).$$

This is equal to a square m^2 if and only if

$$3(2n+1)^2 + 5 = 8m^2.$$

This equation can be rendered in the form $x^2 - 6y^2 = 10$ where $x = 4m$ and $y = 2n + 1$. Solutions to this are

$$(x, y; m, n) = (8, 3; 2, 1), (32, 13; 8, 6), (76, 31; 19, 15),$$

$$(316, 129; 79, 64), \ldots.$$

(c) In particular, we require that $1 + b + b^2$ is a triangular number, i.e., $1 + b + b^2 = (1/2)c(c+1)$, so that

$$(2c+1)^2 - 2(2b+1)^2 = 7.$$

We need to solve $x^2 - 2y^2 = 7$, where $y = 2b + 1$. Solutions are $(x, y; b) = (3, 1; 0)$ (inadmissible), $(5, 3; 1)$, $(13, 9; 4)$, $(27, 19; 9)$, $(75, 53; 26)$, $(157, 111; 55)$, ... Since $1 + b$ is also triangular, the only acceptable value of b in the list so far is $b = 9$.

In fact $b = 9$ works! For each k,

$$(111 \cdots 1)_9 = 1 + 9 + 9^2 + \cdots + 9^{k-1} = \frac{9^k - 1}{8} = \frac{1}{2}r(r+1)$$

with $r = \frac{1}{2}(3^k - 1)$.

Are there any other values of b that will work?

4. (a) Let $a + 1 = y^2$ and $3a + 1 = x^2$. Then $x^2 - 3y^2 = -2$; this has solutions

$$(x, y) = (1, 1), (5, 3), (19, 11), (71, 41), (265, 153), \ldots$$

corresponding to $a = 0, 8, 120, 1680, 23408, \ldots$ respectively. Taking note that $2^2 - 3 \times 1^2 = 1$ and $1^2 - 3 \times 1^2 = -2$, we can construct a sequence of pairs (x_n, y_n) with $x_n^2 - 3y_n^2 = -2$ from the equation

$$x_n + y_n\sqrt{3} = (1 + \sqrt{3})(2 + \sqrt{3})^n.$$

This gives in particular the solutions already noted for $0 \le n \le 4$. Are there any solutions of $x^2 - 3y^2 = -2$ not included in this list?

Suppose that $u^2 - 3v^2 = -2$, where u and v are positive integers with $v > 1$. Observe that u and v must be odd, that $u^2 - v^2 = 2(v^2 - 1) > 0$ so that $u > v$ and that $9v^2 = 6v^2 + u^2 + 2 > u^2$ so that $3v > u$. Hence, both $\frac{1}{2}(3v - u)$ and $\frac{1}{2}(u - v)$ are positive integers.

Since $(1 + \sqrt{3})^{-1} = -\frac{1}{2} + \frac{\sqrt{3}}{2}$,

$$(u + v\sqrt{3})(1 + \sqrt{3})^{-1} = \left(\frac{3v - u}{2}\right) + \left(\frac{u - v}{2}\right)\sqrt{3}.$$

Since

$$\left(\frac{3v - u}{2}\right)^2 - 3\left(\frac{u - v}{2}\right)^2 = \frac{-u^2 + 3v^2}{2} = 1,$$

from the theory of Pell's equation, we must have that

$$\left(\frac{3v - u}{2}\right) + \left(\frac{u - v}{2}\right)\sqrt{3} = (2 + \sqrt{3})^m$$

for some positive integer m, whence we find that $(u, v) = (x_m, y_m)$.

The reader may wish to verify that

$$x_{n+1} = 2x_n + 3y_n \qquad x_{n-1} = 2x_n - 3y_n$$
$$y_{n+1} = x_n + 2y_n \qquad y_{n-1} = -x_n + 2y_n$$
$$x_{n+1} = 4x_n - x_{n-1} \qquad y_{n+1} = 4y_n - y_{n-1}.$$

(b) Let $a_n = y_n^2 - 1$. A little experimentation suggests that $a_n a_{n+1} + 1 = (y_n y_{n+1} - 2)^2$.

Indeed, we find that

$$(y_n y_{n+1} - 2)^2 - (a_n a_{n+1} + 1) = (y_n y_{n+1} - 2)^2 - (y_n^2 - 1)(y_{n+1}^2 - 1) - 1$$
$$= y_n^2 + y_{n+1}^2 - 4y_n y_{n+1} + 2$$
$$= y_n^2 + y_{n+1}(y_{n+1} - 4y_n) + 2$$
$$= y_n^2 - y_{n+1}y_{n-1} + 2$$
$$= y_n^2 - (4y_n - y_{n-1})y_{n-1} + 2$$
$$= y_{n-1}^2 + y_n^2 - 4y_{n-1}y_n + 2$$
$$= (y_{n-1}y_n - 2)^2 - (y_{n-1}^2 - 1)(y_n^2 - 1) - 1.$$

Continuing on, we see ultimately that the difference is equal to

$$(y_0 y_1 - 2)^2 - (a_0 a_1 + 1) = 0,$$

and the result follows.

5. (a) Let n and $n+1$ be the integers and take $x = 2n+1$.

(b) The condition is

$$y^2 = \left(\frac{x-1}{2}\right)^2 + \left(\frac{x+1}{2}\right)^2 = (2x^2+2)/4.$$

(c) $x^2 - 2y^2 = -1$ is satisfied by $(x, y) = (1, 1)$. A complete set of solutions in positive integers is given by

$$x_n + y_n\sqrt{2} = (1 + \sqrt{2})(3 + 2\sqrt{2})^n$$

where n is a positive integer. We find that, for $n \geq 0$,

$$x_{n+1} = 3x_n + 4y_n \quad \text{and} \quad y_{n+1} = 2x_n + 3y_n.$$

[*American Mathematical Monthly* 4 (1897), 24]

6. We have to solve $(u-1)^2 + 2u^2 + 3(u+1)^2 = v^2$ which reduces to $6u^2 + 4u + 4 = v^2$. It is easy to see that v, and thence u, must both be even. Set $u = 2x$ and $v = 2y$ to obtain $(6x+1)^2 - 6y^2 = -5$. Beginning with the obvious solution $(0, 1)$, it is now possible to construct an infinite family of solutions. (Problem 844, *Mathematics Magazine* 45 (1972), 229; 46 (1973), 173–174)

7. See Problem 592, *Mathematics Magazine* 38 (1965), 180; 39 (1966), 75. Since $12(1^5 + 2^5 + \cdots + n^5) = n^2(n+1)^2(2n^2 + 2n - 1)$, the sum of fifth powers is square if and only if $2n^2 + 2n - 1 = 3m^2$ for some integer m. This is equivalent to $(2n+1)^2 - 6m^2 = 3$. Possible values of n include 1, 13, and 133.

8. See the article by W. Golomb in *American Mathematical Monthly* 77 (1970), 848–852, in which a number of problems about powerful numbers are discussed.

(a) If $(x, y) = (u, v)$ satisfies $x^2 - 5y^2 = -1$, then so also does $(x, y) = (9u + 20v, 4u + 9v)$. Starting with $(x, y) = (2, 1)$, we get a chain of solutions that, modulo 5, cycle through the congruence pairs $(2, 1)$, $(3, 2)$, $(2, 0)$, $(3, 3)$, $(2, 4)$, $(3, 4)$, $(2, 3)$, $(3, 0)$, $(2, 2)$, $(3, 1)$. Thus every fifth solution has y a multiple of 5. For example, $682^2 - 5 \times 305^2 = -1$ so that $2^2 \times 11^2 \times 31^2$ and $5^3 \times 61^2$ are consecutive powerful numbers.

(b) An example is $(12167, 12168) = (23^3, 2^3 \times 3^2 \times 13^2)$.

CHAPTER **5**

Equal Sums of Equal Powers

Partitioning the first d^k integers

The first four numbers, 1, 2, 3, 4, can be partitioned into two sets for which the sums of the entries are the same:

$$1 + 4 = 2 + 3.$$

In fact, any four consecutive numbers can be partitioned in this way; simply take the largest with the smallest in one set, and the two middle numbers for the other. Thus, for example, $5 + 8 = 6 + 7$.

Let us check the sum of the squares of the numbers in the partitions:

$$1^2 + 4^2 = 17 \quad \text{and} \quad 2^2 + 3^2 = 13$$

$$5^2 + 8^2 = 89 \quad \text{and} \quad 6^2 + 7^2 = 85.$$

As we can see, the sums of the squares differ. But, for the two examples, the sum of the squares on one side differ from the sum of the squares on the other by the same amount. This means that we can combine the equations to get the following:

$$1 + 4 + 6 + 7 = 2 + 3 + 5 + 8 = 18$$

and

$$1^2 + 4^2 + 6^2 + 7^2 = 2^2 + 3^2 + 5^2 + 8^2 = 102.$$

Thus, we have partitioned the first $8 = 2^3$ numbers into two sets $\{1, 4, 6, 7\}$ and $\{2, 3, 5, 8\}$ for which not only the sums, but also the sums of the squares, of the elements of the two sets are equal. Can this be generalized?

Add the cubes of the numbers in the sets:

$$1^3 + 4^3 + 6^3 + 7^3 = 1 + 64 + 216 + 343 = 624;$$

$$2^3 + 3^3 + 5^3 + 8^3 = 8 + 27 + 125 + 512 = 672.$$

The sums are unequal. But let us increase each of the numbers involved by 8 to get the sets $\{9, 12, 14, 15\}$ and $\{10, 11, 13, 16\}$. The elements of the two sets have the same sum, and it can be checked that the squares of the elements of the sets also have the same sum. Let us examine the sums of the cubes:

$$9^3 + 12^3 + 14^3 + 15^3 = 729 + 1728 + 2744 + 3375 = 8576;$$

and

$$10^3 + 11^3 + 13^3 + 16^3 = 1000 + 1331 + 2197 + 4096 = 8624.$$

Now, observe that $672 - 624 = 8624 - 8576 = 48$, or $672 + 8576 = 8624 + 624$. From this, we conclude that the numbers in the two sets:

$$\{1, 4, 6, 7, 10, 11, 13, 16\} \quad \text{and} \quad \{2, 3, 5, 8, 9, 12, 14, 15\}$$

have, not only the same sums and square sums, but cube sums as well.

As a bonus, we find that for each of the pairs of sets

$$\{1, 4, 6, 7\} \quad \text{and} \quad \{2, 3, 5, 8\}$$
$$\{1, 4, 6, 7, 10, 11, 13, 16\} \quad \text{and} \quad \{2, 3, 5, 8, 9, 12, 14, 15\}$$

the sums of pairs of numbers in the left set are exactly the same (up to repetition) as the corresponding sums for the right set. Does this property continue to hold as we continue the process?

This process can be continued to find the sets

$$\{1, 4, 6, 7, 10, 11, 13, 16, 18, 19, 21, 24, 25, 28, 30, 31\}$$

and

$$\{2, 3, 5, 8, 9, 12, 14, 15, 17, 20, 22, 23, 26, 27, 29, 32\}$$

such that, for each set, the sums, sums of squares, sums of cubes, and sums of fourth powers of the elements are the same as the corresponding sum for the other. Indeed, given any positive integer k, we can partition the two set $\{1, 2, 3, 4, 5, \ldots, 2^{k+1}\}$ into two subsets whose powers sums are equal from the first up to the kth power.

This fact can be generalized in another way. Let us partition the set

$$\{1, 2, 3, 4, 5, 6, 7, 8, 9\}$$

into three subsets whose elements have the same sum. Since the sum of all nine integers is 45, we want the sum of the elements of each subset to be 15. We can get an indication of how to do this by recalling the magic square

$$
\begin{array}{ccc}
4 & 3 & 8 \\
9 & 5 & 1 \\
2 & 7 & 6.
\end{array}
$$

One possible partition is $\{1, 6, 8\}$, $\{2, 4, 9\}$, $\{3, 5, 7\}$, suggested by the columns of the magic square. The sums of the squares of the elements of these three sets are, respectively, 101, 101, 83. If we add 9 to each number in these sets to get new ones, we find that the square sums are 614, 614, 596. Adding 18 gives sets with square sums 1613, 1613, 1595. By patching the sets together, we can get a partition of the first 27 positive integers into three sets for which the sums and square sums of the numbers are equal. For example, we can take

$$\{1, 6, 8, 12, 14, 16, 20, 22, 27\}$$

$$\{2, 4, 9, 10, 15, 17, 21, 23, 25\}$$

$$\{3, 5, 7, 11, 13, 18, 19, 24, 26\}.$$

This is not the only possibility. Taking the partition of the first nine integers suggested by the rows of the magic square, we eventually can arrive at the partition

$$\{1, 5, 9, 12, 13, 17, 20, 24, 25\}$$

$$\{3, 4, 8, 11, 15, 16, 19, 23, 27\}$$

$$\{2, 6, 7, 10, 14, 18, 21, 22, 26\}.$$

The reader is invited to partition the numbers from 1 to 81 inclusive into three sets for which the sums, square sums, and cube sums are all the same. Then you might try the following problems:

1. partition the numbers from 1 to 64 inclusive into four sets with the same sums and square sums;
2. partition the numbers from 1 to 256 inclusive into four sets with the same sums, square sums, and cube sums;
3. partition the numbers from 1 to 125 inclusive into five sets with the same sum and square sums.

What is the ultimate generalization to all of this? **(1)**

In a different direction, we have the interesting result that there are n consecutive integers whose squares can be divided into two sets having the same sum if and only if n can be expressed as the difference of two positive squares **(2)**. This condition is equivalent to n being unequal to 1, 4 or any number of the form $4k + 2$.

For example, if $n = 3 = 2^2 - 1^2$, then $3^2 + 4^2 = 5^2$; if $n = 5 = 3^2 - 2^2$, then $2^2 + 4^2 + 5^2 = 3^2 + 6^2$; and if $n = 7 = 4^2 - 3^2$, then $5^2 + 7^2 + 8^2 + 10^2 = 6^2 + 9^2 + 11^2$. The case $n = 8k$ is covered by the discussion at the beginning of this chapter. The squares $1^2, 2^2, \ldots, (8k)^2$ can be subdivided into two subsets having the same sum according as to whether the roots leaves remainders in the set $\{1, 4, 6, 7\}$ or the set $\{2, 3, 5, 8\}$ upon division by 8.

When $n = 8k + 4$, then again we take the first n squares and put in one set the squares of 6, 8 and all integers exceeding 8 leaving remainders 1, 4, 6 or 7 upon division by 8. For example, when $n = 20$, we have that

$$1^2 + 2^2 + 3^2 + 4^2 + 5^2 + 7^2 + 10^2 + 11^2 + 13^2 + 16^2 + 18^2 + 19^2$$
$$= 6^2 + 8^2 + 9^2 + 12^2 + 14^2 + 15^2 + 17^2 + 20^2.$$

For odd values of n, we have to start the n squares higher up. When $n = 4k + 1$, we take n consecutive squares starting with that of $2k$, putting in one of the sets the smallest k odd and the largest k even squares. When $n = 4k + 3$, we take the n squares starting with that of $2k + 3$, putting in one set the smallest $k + 1$ odd squares and the largest $k + 1$ even squares.

The Tarry–Escott Problem

The process which we followed to partition a set of numbers into two subsets whose various power sums are equal can be strengthened to give two sets as small as possible with equal power sums. Suppose we ask for two nonoverlapping sets of numbers, each with m elements whose power sums from the first up to the nth are equal. In the first section, we saw how to do this with $m = 2^{k+1}$ and $n = k$, for positive integer k. It is known that m must be strictly greater than n. The Tarry–Escott problem is to make $m = n + 1$, i.e., for any positive integer n, to construct two sets each with $n + 1$ integers, for which the sums, sums of squares, sums of cubes, up to the sums of nth powers of the two sets are respectively equal.

For two sets of numbers, A and B, let us write $A =^k B$ to denote the fact that each of the power sums from the first up to the kth power of the two sets are equal. We can start with simple relations and build up to complicated relations using the sort of shifting by adding constants to the sets as we did in the first section. The reader is invited to investigate how each equation in the following chain can be found from its predecessors **(3)**:

$$\{1, 4\} =^1 \{2, 3\}$$
$$\{4, 7\} =^1 \{5, 6\}$$
$$\{1, 4, 5, 6\} =^2 \{2, 3, 4, 7\}$$
$$\{1, 5, 6\} =^2 \{2, 3, 7\}$$
$$\{6, 10, 11\} =^2 \{7, 8, 12\}$$

$$\{1, 5, 6, 7, 8, 12\} =^3 \{2, 3, 7, 6, 10, 11\}$$

$$\{1, 5, 8, 12\} =^3 \{2, 3, 10, 11\}$$

$$\{8, 12, 15, 19\} =^3 \{9, 10, 17, 18\}$$

$$\{1, 5, 8, 9, 10, 12, 17, 18\} =^4 \{2, 3, 8, 10, 11, 12, 15, 19\}$$

$$\{1, 5, 9, 17, 18\} =^4 \{2, 3, 11, 15, 19\}$$

$$\{9, 13, 17, 25, 26\} =^4 \{10, 11, 19, 23, 27\}$$

$$\{1, 5, 10, 18, 23, 27\} =^5 \{2, 3, 13, 15, 25, 26\}$$

$$\{14, 18, 23, 31, 36, 40\} =^5 \{15, 16, 26, 28, 38, 39\}$$

$$\{1, 5, 10, 16, 27, 28, 38, 39\} =^6 \{2, 3, 13, 14, 25, 31, 36, 40\}$$

$$\{12, 16, 21, 27, 38, 39, 49, 50\} =^6 \{13, 14, 24, 25, 36, 42, 47, 51\}$$

$$\{1, 5, 10, 24, 28, 42, 47, 51\} =^7 \{2, 3, 12, 21, 31, 40, 49, 50\}.$$

The reader may wish to check that, in the last of these, the sums of the first, second, up to the seventh powers of the elements on one side are respectively equal to the corresponding sums on the other.

There is a slightly easier way to do this. If you have been doing some experimentation, you will realize that the relation $(=^k)$ of power sum equality up to the kth power between two sets is preserved if you multiply each number in the sets by the same constant, when you remove numbers common to both sets and when you add to both sides or subtract from both sides the same constant. Let us look at the last equation in the foregoing list, and "center it up" by subtracting 26 from each of the numbers involved. The last equation is true exactly when the following is true:

$$\{-25, -21, -16, -2, 2, 16, 21, 25\} =^7 \{-24, -23, -14, -5, 5, 14, 23, 24\}.$$

Since each of the sets contains a number only if the negative of that number is included, it is clear that the sums of any odd power of the numbers in the two sets are both zero. Thus, to check the equation, it suffices only to check that

$$2^2 + 16^2 + 21^2 + 25^2 = 5^2 + 14^2 + 23^2 + 24^2$$

$$2^4 + 16^4 + 21^4 + 25^4 = 5^4 + 14^4 + 23^4 + 24^4$$

$$2^6 + 16^6 + 21^6 + 25^6 = 5^6 + 14^6 + 23^6 + 24^6.$$

This task is left to the reader. As a bonus, we observe that the two sets $\{-2, 16, -21, 25\}$ and $\{5, 14, 23, -24\}$ have four equal power sums: the first, second, fourth, and sixth.

There are many more such sets to be discovered. For example, we have

$$\{0, 5, 6, 16, 17, 22\} =^5 \{1, 2, 10, 12, 20, 21\}$$

$$\{0, 18, 27, 58, 64, 89, 101\} =^6 \{1, 13, 38, 44, 75, 84, 102\}$$

$$\{1, 13, 28, 70, 82, 124, 139, 151\} =^7 \{4, 7, 34, 61, 91, 118, 145, 148\}$$

$$\{-37, -35, -15, -13, 9, 21, 31, 39\} =^6 \{-39, -31, -21, -9, 13, 15, 35, 37\}$$

$$\{-67, -65, -47, -31, -13, 15, 27, 51, 61, 69\} =^8$$

$$\{-69, -61, -51, -27, -15, 13, 31, 47, 65, 67\}$$

$$\{-57, -55, -33, -23, 1, 5, 13, 39, 51, 59\} =^8$$

$$\{-59, -51, -39, -13, -5, -1, 23, 33, 55, 57\}$$

and the parametric ones

$$\{a, a + 4b + c, a + b + 2c, a + 9b + 4c, a + 6b + 5c, a + 10b + 6c\} =^5$$

$$\{a + b, a + c, a + 6b + 2c, a + 4b + 4c, a + 10b + 5c, a + 9b + 6c\};$$

$$\{-5x - 3y, -4x - y, -x - 2y, x + 2y, 4x + y, 5x + 3y\} =^5$$

$$\{-4x - 3y, -5x - 2y, x - y, -x + y, 5x + 2y, 4x + 3y\};$$

and **(4)**

$$\{0, a^2 + 2ab + 6b^2, 2a^2 - 7ab + b^2, 3a^2 + 8ab + 7b^2,$$

$$5a^2 - 5ab - 3b^2, 6a^2 + 10ab + 5b^2, 7a^2 + ab - 2b^2, 8a^2 + 3ab + 4b^2\} =^7$$

$$\{ab + 5b^2, a^2 - 4ab - 3b^2, 2a^2 + 7ab + 6b^2, 3a^2 - 6ab, 5a^2 + 9ab + 4b^2,$$

$$6a^2 - 4ab - 2b^2, 7a^2 + 7ab + 7b^2, 8a^2 + 2ab - b^2\}.$$

The last three numerical examples have as a byproduct the equations

$$9^k + 21^k + 31^k + 39^k = 13^k + 15^k + 35^k + 37^k \quad \text{for } k = 1, 3, 5$$

$$13^k + 31^k + 47^k + 65^k + 67^k = 15^k + 27^k + 51^k + 61^k + 69^k \quad \text{for } k = 1, 3, 5, 7$$

$$1^k + 5^k + 13^k + 39^k + 51^k + 59^k = 23^k + 33^k + 55^k + 57^k \quad \text{for } k = 1, 3, 5, 7.$$

How high can you make n for which you can find two sets of $n + 1$ numbers which satisfy the $=^n$ equality?

Sums and square sums

We look at some unusual ways of finding pairs of sets of numbers whose sums and square sums are equal. The first involves the magic square **(5)**.

Take the three digit numbers suggested by the rows of the square: 438, 951, 276. Form a second set by reversing the digits. We find that

$$\{438, 951, 276\} =^2 \{834, 159, 672\}.$$

Doing a similar thing with the columns yields

$$\{492, 357, 816\} =^2 \{294, 753, 618\}.$$

Not to be outdone, the diagonal directions, both ways, yield

$$\{456, 231, 978\} =^2 \{654, 132, 879\}$$

and

$$\{852, 174, 639\} =^2 \{258, 471, 936\}.$$

For an alternative way **(6)**, form any $n \times n$ square array (a_{ij}) of 0s and 1s for which $a_{ii} = 0$ and $a_{ij} + a_{ji} = 1$ $(1 \le i, j \le n)$. This means that the diagonal elements are 0 and, for the nondiagonal elements, the reflection about the diagonal of a 0 is a 1. An example is

$$
\begin{array}{ccccccc}
0 & 1 & 1 & 1 & 1 & 1 & 0 \\
0 & 0 & 1 & 1 & 0 & 1 & 1 \\
0 & 0 & 0 & 1 & 1 & 0 & 0 \\
0 & 0 & 0 & 0 & 0 & 1 & 1 \\
0 & 1 & 0 & 1 & 0 & 0 & 1 \\
0 & 0 & 1 & 0 & 1 & 0 & 1 \\
1 & 0 & 1 & 0 & 0 & 0 & 0 \\
\end{array}
$$

Let $\{r_1, r_2, \ldots, r_n\}$ be the set of row sums and $\{c_1, c_2, \ldots, c_n\}$ be the set of column sums. Then, it always happens that

$$\{r_1, r_2, \ldots, r_n\} =^2 \{c_1, c_2, \ldots, c_n\}.$$

For the example, we have that

$$\{5, 4, 2, 2, 3, 3, 2\} =^2 \{1, 2, 4, 4, 3, 3, 4\}.$$

The equations

$$3^3 + 4^3 + 5^3 + 6^3 = 91 + 152 + 189$$

and

$$3^6 + 4^6 + 5^6 + 6^6 = 91^2 + 152^2 + 189^2$$

are a restatement of the relation

$$\{3^3, 4^3, 5^3, 6^3\} =^2 \{0, 91, 152, 189\}.$$

The numbers involved are connected in other ways: $3^3 + 4^3 = 91$; $3^3 + 5^3 = 152$; $4^3 + 5^3 = 189$; $3^3 + 4^3 + 5^3 = 6^3$. This gives a clue for a method of constructing other examples **(7)**. Select any four numbers a, b, c, d for which $a^3 + b^3 + c^3 = d^3$. Let

$$u = b^3 + c^3; \quad v = a^3 + c^3; \quad w = a^3 + b^3.$$

Then

$$u + v + w = 2(a^3 + b^3 + c^3) = a^3 + b^3 + c^3 + d^3$$

and

$$
\begin{aligned}
u^2 + v^2 + w^2 &= 2(a^6 + b^6 + c^6 + b^3 c^3 + a^3 c^3 + a^3 b^3) \\
&= a^6 + b^6 + c^6 + (a^6 + b^6 + c^6 + 2a^3 b^3 + 2a^3 c^3 + 2b^3 c^3) \\
&= a^6 + b^6 + c^6 + d^6.
\end{aligned}
$$

For example, beginning with $1^3 + 6^3 + 8^3 = 9^3$, we find that

$$\{1^3, 6^3, 8^3, 9^3\} =^2 \{0, 217, 513, 728\}.$$

A 4×4 magic square figures in a famous woodcut entitled *Melancholia*, executed in 1514 by the artist Albrecht Dürer (1471–1528). This square

16	3	2	13
5	10	11	8
9	6	7	12
4	15	14	1

has some interesting properties involving squares and cubes of its entries **(8)**. Not only the sums of the numbers in each row, in each column, and in each diagonal equal to 34, but looking at the entries in the first two and the last two rows, we find that

$$\{16, 3, 2, 13, 5, 10, 11, 8\} =^2 \{9, 6, 7, 12, 4, 15, 14, 1\}.$$

From alternate columns, we discover the relation

$$\{16, 5, 9, 4, 2, 11, 7, 14\} =^2 \{3, 10, 6, 15, 13, 8, 12, 1\}.$$

The sums of the powers up to the third are equal for the sets of numbers on and off the diagonals. Indeed, we have the example already indicated at the outset of this chapter

$$\{16, 10, 7, 1, 13, 11, 6, 4\} =^3 \{2, 3, 5, 8, 9, 12, 14, 15\}.$$

Equal products and square sums

A variant on what has just been discussed is to ask for the *product* and square sums of two sets to be equal. An example is the pair of sets, $\{3, 8, 26\}$ and $\{2, 13, 24\}$. The product of the numbers in each set is $2 \times 3 \times 8 \times 13$ and the sum of the squares is 749. Are there other pairs of sets $\{x, y, z\}$ and $\{u, v, w\}$ for which **(9)**

$$x^2 + y^2 + z^2 = u^2 + v^2 + w^2$$

while

$$xyz = uvw?$$

There is a nice method of generating them, due to E. T. Bell **(10)**. Consider the matrix

$$\begin{pmatrix} a & b & c \\ d & e & f \\ g & h & k \end{pmatrix}.$$

We try $x = abc$, $y = def$, $z = ghk$ (the row products) and $u = adg$, $v = beh$, $w = cfk$ (the column products). This will work if we choose a, b, c, d, e, f, g, h, k to satisfy the side condition:

$$a^2(b^2c^2 - d^2g^2) + e^2(d^2f^2 - b^2h^2) + k^2(g^2h^2 - c^2f^2) = 0.$$

Suitable matrices are

$$\begin{pmatrix} 1 & 2q-p & p-q \\ p & 1 & 2p-q \\ p+q & q & 1 \end{pmatrix},$$

$$\begin{pmatrix} 1 & 1 & 2p+1 \\ 1 & 1 & 3p-2 \\ p & 6p-1 & 1 \end{pmatrix}, \quad \text{and} \quad \begin{pmatrix} 1 & 1 & 2p-1 \\ 1 & 1 & 3p+2 \\ p & 6p+1 & 1 \end{pmatrix}$$

where p and q are integers.

A parametric solution to $xyz = uvw$ and $x^2 + y^2 + z^2 = u^2 + v^2 + w^2$ is given by **(11)**

$$(x, y, z) = \big(b(c^2 - b^2 + ac), (a+c)(b^2 - c^2 + ac), ab(a-c)\big)$$

$$(u, v, w) = \big(b(b^2 - c^2 + ac), (a-c)(c^2 - b^2 + ac), ab(a+c)\big).$$

Exercises on the Notes

2. *Straightforward high school algebra.* (a) Verify the following two identities and use them to check the assertions about the numbers $n = 8k$ and

$8k + 4$:

$$(8x + 1)^2 + (8x + 4)^2 + (8x + 6)^2 + (8x + 7)^2$$
$$= (8x + 2)^2 + (8x + 3)^2 + (8x + 5)^2 + (8x + 8)^2$$
$$6^2 + 8^2 + (8x + 1)^2 + (8x + 4)^2$$
$$= 1^2 + 2^2 + 3^2 + 4^2 + 5^2 + 7^2 + (8x + 2)^2 + (8x + 3)^2.$$

(b) Verify that

$$\sum_{i=1}^{k}(2k + 2i - 1)^2 + \sum_{i=1}^{k}(4k + 2i)^2 = \sum_{i=0}^{k}(2k + 2i)^2 + \sum_{i=1}^{k}(4k + 2i - 1)^2.$$

(c) Verify that

$$\sum_{i=1}^{k+1}(2k+2i+1)^2 + \sum_{i=1}^{k+1}(4k+2i+2)^2 = \sum_{i=1}^{k}(2k+2i+2)^2 + \sum_{i=1}^{k+1}(4k+2i+3)^2.$$

10. *Straightforward high school algebra.* Verify the parametric solutions. (In the case of the sums of squares, an efficient approach is to look at

$$(x^2 - u^2) + (y^2 - v^2) + (z^2 - w^2).)$$

Notes

1. If d and m are any positive integers with $d \geq 2$, the set of numbers from 1 to d^{m+1} inclusive can be subdivided into d disjoint subsets in such a way that, for any polynomial of degree not exceeding m, the sum of its values over any of the subsets is the same. Write each number in the set $\{0, 1, 2, \ldots, d^{m+1} - 1\}$ to base d, sum the digits, and classify according to the congruence of that sum modulo d; then add 1 to each number to get the sets.

　　For example, the numbers from 0 to 26, in base 3, can be subdivided into the sets

$$\{0, 12, 21, 102, 111, 120, 201, 210, 222\},$$
$$\{1, 10, 22, 100, 112, 121, 202, 211, 220\},$$
$$\{2, 11, 20, 101, 110, 122, 200, 212, 221\}.$$

This yields the first splitting of $\{1, 2, \ldots, 27\}$ given in the text.

2. Problem E1972, *American Mathematical Monthly* 74 (1967), 318; 75 (1968), 679–680.

3. Suppose that $\{a_1, a_2, \ldots, a_r\} =^n \{b_1, b_2, \ldots, b_r\}$. It is straightforward to show that for any polynomial $p(t) = c_n t^n + c_{n-1} t^{n-1} + \cdots + c_1 t + c_0$ of degree not exceeding n,

$$\sum_{i=1}^{r} p(a_i) = \sum_{i=1}^{r} p(b_i).$$

Also, $\sum_{i=1}^{r} p(ka_i) = \sum_{i=1}^{r} p(kb_i)$ and

$$\sum_{i=1}^{r} p(a_i + k) = \sum_{i=1}^{r} \sum_{j=0}^{n} \frac{k^j}{j!} p^{(j)}(a_i) = \sum_{j=0}^{n} \frac{k^j}{j!} \sum_{i=1}^{r} p^{(j)}(a_i)$$

$$\sum_{j=0}^{n} \frac{k^j}{j!} \sum_{i=1}^{r} p^{(j)}(b_i) = \sum_{i=1}^{r} p(b_i + k)$$

whence

$$\{ka_1, ka_2, \ldots, ka_r\} =^n \{kb_1, kb_2, \ldots, kb_r\}$$

and

$$\{a_1 + k, a_2 + k, \ldots, a_r + k\} =^n \{b_1 + k, b_2 + k, \ldots, b_r + k\}$$

for any constant k.

Since

$$\sum_{i=1}^{r} (a_i + k)^{n+1} - \sum_{i=1}^{r} (b_i + k)^{n+1}$$

$$= \sum_{i=1}^{r} (a_i^{n+1} - b_i^{n+1}) + \sum_{j=1}^{n+1} \binom{n+1}{j} k^j \left[\sum_{i=1}^{r} a_i^{n+1-j} - \sum_{i=1}^{r} b_i^{n+1-j} \right]$$

$$= \sum_{i=1}^{r} a_i^{n+1} - \sum_{i=1}^{r} b_i^{n+1},$$

the difference between the sum of the $(n+1)$th powers for the left and right sets remains unchanged when the same constant is added to every number in both sets.

4. This was given by D. Cross in *Journal of Recreational Mathematics* 11 (1978–1979), 42.

5. To understand what it going on, we draw attention to some striking properties of the magic square

$$
\begin{array}{ccc}
4 & 3 & 8 \\
9 & 5 & 1 \\
2 & 7 & 6
\end{array}
$$

The respective sums of the squares of the row elements are 89, 107, 89 and the corresponding sums for the columns are 101, 83, 101. We also have that

$$4 \times 9 + 3 \times 5 + 8 \times 1 = 59 = 9 \times 2 + 5 \times 7 + 1 \times 6$$

$$4 \times 3 + 9 \times 5 + 2 \times 7 = 71 = 3 \times 8 + 5 \times 1 + 7 \times 6$$

$$4 \times 5 + 9 \times 7 + 2 \times 3 = 89 = 5 \times 6 + 7 \times 8 + 3 \times 1$$

$$8 \times 5 + 1 \times 7 + 6 \times 3 = 65 = 5 \times 2 + 7 \times 4 + 3 \times 9$$

Using these facts, it is straightforward to see that for $k = 1$ and $k = 2$,

$$(4x^2 + 3x + 8)^k + (9x^2 + 5x + 1)^k + (2x^2 + 7x + 6)^k$$
$$= (8x^2 + 3x + 4)^k + (x^2 + 5x + 9)^k + (6x^2 + 7x + 2)^k$$
$$(4x^2 + 9x + 2)^k + (3x^2 + 5x + 7)^k + (8x^2 + x + 6)^k$$
$$= (2x^2 + 9x + 4)^k + (7x^2 + 5x + 3)^k + (6x^2 + x + 8)^k$$
$$(4x^2 + 5x + 6)^k + (2x^2 + 3x + 1)^k + (9x^2 + 7x + 8)^k$$
$$= (6x^2 + 5x + 4)^k + (x^2 + 3x + 2)^k + (8x^2 + 7x + 9)^k$$
$$(8x^2 + 5x^2 + 2)^k + (x^2 + 7x + 4)^k + (6x^2 + 3x + 9)^k$$
$$= (2x^2 + 5x + 8)^k + (4x^2 + 7x + 1)^k + (9x^2 + 3x + 6)^k.$$

Setting $x = 10$ yields the relations given. Note that we can take for x any base of enumeration exceeding 10.

In Notes 2963 and 3149 of *Mathematical Gazette* 45 (1961), 224–227; 50 (1966), 173–175, D. C. Cross draws attention to equations involving squares of elements in various magic squares.

6. This property of matrices was noted by Nick Mackinnon in *Mathematical Gazette* 73 (1989), 134 (item 73.24). The sum of the row sums equals the sum of the column sums equals the number of ones in the array. The proof that the square sums are equal depends upon two facts.

(i) If U is the upper triangular $n \times n$ matrix defined by $u_{ij} = 1$ if $i < j$ and $u_{ij} = 0$ if $j \geq i$, then the sets of row and column sums are both $\{1, 2, \ldots, (n - 1)\}$ and the result holds for U.

(ii) Suppose the result holds for a matrix $A = (a_{ij})$ and that the matrix $B = (b_{ij})$ is obtained from A by interchanging the (p, q)th and (q, p)th element for some $p \neq q$:

$$a_{ij} = b_{ij} \quad \text{for } (i, j) \neq (p, q), (q, p)$$

$$a_{pq} = b_{qp} \quad \text{and} \quad a_{qp} = b_{pq}.$$

Then the result holds for B. To see this, let r_i and c_j denote the row and column sums for A. Then supposing without loss of generality that $a_{pq} = 0 = b_{qp}$, the sum of the

pth row of B is $r_p + 1$
qth row of B is $r_q - 1$
pth column of B is $c_p - 1$
qth column of B is $c_q + 1$.

The difference between the sum of the squares of the row sums and the sum of squares of the column sums for B is

$$\left[\sum_{i \neq p, q} r_i^2 + (r_p + 1)^2 + (r_q - 1)^2 \right] - \left[\sum_{i \neq p, q} c_i^2 + (c_p - 1)^2 + (c_q + 1)^2 \right]$$

$$= 2(r_p + c_p - r_q - c_q).$$

Since $r_i + c_i = n - 1$ for each i, this difference is 0.

To conclude the argument, simply observe that each acceptable matrix can be obtained from U by performing a sequence of switches described in (ii).

7.   See Problem 3870 in *American Mathematical Monthly* 45 (1938), 253.

8.   See pages 211–213 of Howard Eves, *An introduction to the history of mathematics,* Holt, Rinehart, Winston. 5th edition, 1983. This fact was the basis of the calendar problem for March 16 in *Mathematics Teacher* 88 (1995), 219.

9.   In *American Mathematical Monthly* 95 (1988), 35, a systematic method of producing such pairs is sought. The problem is taken up in a later paper by J. B. Kelly in *American Mathematical Monthly* 98 (1991), 527–529.

10.   *American Journal of Mathematics* 55 (1933), 50–66.

11.   *American Mathematical Monthly* 75 (1968), 1061–1073 (Section 3.3).

References

There are many references in the literature to the problem of finding distinct sets which have equal sums of several powers. The determination of two disjoint sets of $n + 1$ integers all of whose power sums from the first up to the nth are equal is often called the *Tarry–Escott* problem. The most thorough discussion occurs in the monograph of Albert Gloden, listed below. A history of the problem can be found in Dickson's *History,* and there is a section on it in Hardy and Wright.

I. A. Barnett and C. W. Mendel, On equal sums of squares, *American Mathematical Monthly* 49 (1942), 157–170.

Ian Barrodale, A note on equal sums of like powers, *Mathematics of Computation* 20 (1966), 318–322.

Albert H. Beiler, *Recreations in the theory of number—the queen of mathematics entertains,* Dover, p. 162.

J. Chernick, Ideal solutions of the Tarry–Escott problem, *American Mathematical Monthly* 44 (1937), 626–633.

Donald Cross, Second- and third-order multi-multigrades, *Journal of Recreational Mathematics* 7 (1974), 41–44.

——, Third-order multi-multigrades from sums of two squares, *Journal of Recreational Mathematics* 7 (1974), 183–185.

——, Fifth order multi-multigrades, *Journal of Recreational Mathematics* 8 (1975–1976), 239–240.

——, Letter to the editor, *Journal of Recreational Mathematics* 11 (1978–1979), 42.

L. E. Dickson, *History of the theory of numbers,* Vol. II. (Washington, 1920; reprint Chelsea, 1952) Chapter 24.

——, *Introduction to the theory of numbers,* Chicago, 1929; Dover, 1957. See *Sets of integers having equal sums of like powers,* pages 49–58.

H. L. Dorwart, Sequences of ideal solutions in the Tarry–Escott problem, *Bulletin of the American Mathematical Society* 53 (1947), 381–391.

——, Old fashioned algebra can be useful, *Mathematics Magazine* 54 (1981), 11–13.

H. L. Dorwart and O. E. Brown, The Tarry–Escott problem, *American Mathematical Monthly* 44 (1937), 613–626.

Harold L. Dorwart and Warren Page, Introduction to the Tarry–Escott problem, *Two-year College Mathematics Readings* (1981), 87–95.

E. B. Escott, Logarithmic series, *Quarterly Journal of Pure and Applied Mathematics* 41 (1910), 141–167.

A. Gloden, *Mehrgradige Gleichungen,* (Noordhoff, Groningen, 1944) **MR 8,** 441f.

——, Two theorems on multi-degree equalities, *American Mathematical Monthly* 53 (1946), 205–206.

——, Parametric solutions of two multi-degree equalities, *American Mathematical Monthly* 55 (1948), 86–88.

——, Normal trigrade and cyclic quadrilaterals with integral sides and diagonals, *American Mathematical Monthly* 58 (1951), 244–247.

G. H. Hardy and E. M. Wright, *An introduction to the theory of numbers,* 4th edition, (Oxford, 1960) pp. 328–332.

Maurice Kraitchik, *Mathematical recreations,* (Norton, 1942; Dover) p. 79.

Loren Larson, *Problem solving through problems,* (Springer, 1983) Problem 5.1.15; pp. 163–164.

D. H. Lehmer, The Tarry–Escott problem, *Scripta Math.* 13 (1947), 37–41.

Joseph S. Madachy, *Mathematics on vacation* (Scribner's, 1966) pp. 173–175.

Gerald Myerson, How small can a sum of roots be? *American Mathematical Monthly* 93 (1986), 457–459.

J. B. Roberts, Λ curious sequence of signs, *American Mathematical Monthly* 64 (1957), 317–322.

——, Splitting consecutive integers into classes with equal power sums, *American Mathematical Monthly* 71 (1964), 25–37.

Joe Roberts, *Elementary number theory: a problem oriented approach,* MIT, 1977 Pages 88, 110S–111S.

——, *Lure of the integers,* MAA Spectrum, 1992, p. 24.

T. N. Sinha, On the Tarry–Escott problem, *American Mathematical Monthly* 73 (1966), 280–285.

G. Tarry, *L'Intermédiaire des Mathématiciens* 19 (1912), 219–221.

J. S. Vidger, Consecutive integers having equal sums of squares, *Mathematics Magazine* 38 (1965), 35–42.

E. M. Wright, On Tarry's problem, *Quarterly J. Mathematics Oxford* (1) 6 (1935), 261–267; 7 (1936), 43–45; 8 (1937), 48–50.

——, Prouhet's 1851 solution of the Tarry–Escott problem of 1910, *American Mathematical Monthly* 66 (1959), 199–201.

Number curiosities, *Crux Mathematicorum (Eureka)* 2 (1976), 62.

Problem 963, *Crux Mathematicorum* 10 (1984), 216; 11 (1985), 292–296.

Problem E332, *American Mathematical Monthly* 45 (1938), 249; 46 (1939), 172–173.

Problem E1312, *American Mathematical Monthly* 65 (1958), 284, 776.

Problem E1504, *American Mathematical Monthly* 69 (1962), 165, 924.

Problem 10284, *American Mathematical Monthly* 100 (1993), 185; 102 (1995), 843–844.

Exercises

1. (a) Observe that $\{1, 4, 4\} =^2 \{5, 2, 2\}$ and that $1 + 5 = 4 + 2 = 6$. Determine all quadruples of positive integers $(a, b, c; k)$ for which $\{a, b, c\} =^2 \{k - a, k - b, k - c\}$.

    (b) Find a generalization of (a).

2. Establish the identity

$$\sum_{r=0}^{n-2} (2^r - 1)^2 + \left[3(2^{n-1}) - 1 \right]^2 = \sum_{r=2}^{n} (2^r + 1)^2 + (2^n - 4)^2,$$

thus obtaining a general method for finding two equal sums of n distinct squares.

Solutions

1. (a) Since $a + b + c = (k - a) + (k - b) + (k - c)$, $2(a + b + c) = 3k$. Then $(a^2 + b^2 + c^2) - ((k - a)^2 + (k - b)^2 + (k - c)^2) = 2k(a + b + c) - 3k^2 = 0$. So, all that is required is to select a, b, and c so that their sum is divisible by 3 and take $k = (2/3)(a + b + c)$. For example, $\{3, 5, 10\} =^2 \{9, 7, 2\}$.

    (b) A generalization of (a) is

    if $a_1, a_2, \ldots, a_n$ are positive integers whose sum is divisible by n and if $k = 2(a_1 + a_2 + \cdots + a_n)/n$, then

$$\{a_1, a_2, \ldots, a_n\} =^2 \{k - a_1, k - a_2, \ldots, k - a_n\}.$$

Remarks. (1) This gives an alternative proof for item **(6)**. In the matrix, $r_1 + r_2 + \cdots + r_n = n(n - 1)/2$ and $c_i = (n - 1) - r_i$ for each i.

    (2) In Volume II of his *History*, Dickson reports on earlier formulations of this result by Catalan (p. 295) and Euler (p. 261). Euler also noted that

$$a^2 + b^2 + c^2 = (k - a)^2 + (k - b)^2 + (2k - c)^2$$

when $a + b + 2c = 3k$, and

$$a^2 + b^2 + c^2 = (2k - a)^2 + (4k - b)^2 + (4k - c)^2$$

when $a + 2b + 2c = 9k$.

2. See Problem E1286 in *American Mathematical Monthly* 64 (1957), 670; 65 (1958), 366–367.

$$\sum_{r=2}^{n} (2^r + 1)^2 - \sum_{r=0}^{n-2} (2^r - 1)^2 = \sum_{r=0}^{n-2} \left[(4 \cdot 2^r + 1)^2 - (2^r - 1)^2 \right]$$

$$= 5 \sum_{r=0}^{n-2} (3 \cdot 2^{2r} + 2^{r+1})$$

$$= 5(2^{2n-2} - 1 + 2^n - 2)$$

$$= (5 \cdot 2^{n-1} - 5)(2^{n-1} + 3)$$

$$= [3 \cdot 2^{n-1} - 1]^2 - (2^n - 4)^2.$$

CHAPTER 6

Digits and Sums of Powers

Numbers which are sums of powers of their digits

The numbers 153, 370, 371, and 407 all have the interesting property that they are the sums of the cubes of their digits. For example, $371 = 27 + 343 + 1 = 3^3 + 7^3 + 1^3$. These four are the only numbers, apart from the trivial case, 1, with this property. There are also numbers that are sums of higher powers of their digits **(1)**. Thus we have the following table:

k	Numbers equal to the sum of the kth power of their digits
3	$153, 370, 371, 407$
4	$1634, 8208, 9474$
5	$4150, 4151, 54748, 92727, 93084, 194979$
6	548834
7	$1741725, 4210818, 9800817, 9926315, 14459929$
8	$24678050, 24678051, 88593477$
9	$146511208, 472335975, 534494836, 912985153$

If we ask for numbers besides 1 which are the sums of the squares of their digits, we will be disappointed, as there are none **(2)**.

However, if we turn to other bases, we get an affirmative answer. Thus, in base 3, we find that

$$1 \times 3 + 2 = (12)_3 = 1^2 + 2^2 = 5; \quad 2 \times 3 + 2 = (22)_3 = 2^2 + 2^2 = 8$$

and in base 5, that

$$2 \times 5 + 3 = (23)_5 = 2^2 + 3^2 = 13; \quad 3 \times 5 + 3 = (33)_5 = 3^2 + 3^2 = 18.$$

In general, if the base b is the odd number $2k + 1$, we have

$$k \times (2k + 1) + (k + 1) = 2k^2 + 2k + 1 = k^2 + (k + 1)^2$$

and

$$(k+1) \times (2k+1) + (k+1) = 2k^2 + 4k + 2 = (k+1)^2 + (k+1)^2.$$

We see that, whatever odd base is given, we can find numbers, which when written to that base, are the sums of the squares of its digits.

Any number to any base, except 1, which is equal to the sum of the squares of its digits, must have two digits. Thus, such numbers can be written as the sum of two squares. If we want to find further examples, we pick numbers expressible as the sum of two squares and then cook up a base for which it is the sum of its digits.

For example, $25 = 3^2 + 4^2$. Can we find a base b for which the number 25 written to that base has digits 3 and 4. That is, we require, either $25 = 3b + 4$ or $25 = 4b + 3$. The first of these is solvable for a positive integer $b = 7$ and we obtain an example discussed earlier. The second of these does not yield a positive integer b. To approach the equation in a systematic way, we can look at numbers which are the sums of two squares and find appropriate bases. A table might begin:

Number	Sum of the squares	Possible bases
5	1, 2	3
10	1, 3	7
13	2, 3	5
17	1, 4	13
20	2, 4	8
25	3, 4	7
26	1, 5	21
	. . .	
65	1, 8	57
65	4, 7	none
	. . .	

If the base is a power of 10, then we get some striking numerical relations (3):

$$12 \times 100 + 33 = 1233 = 12^2 + 33^2$$

$$588 \times 10000 + 2353 = 5882353 = 588^2 + 2353^2$$

$$990 \times 100000 + 09901 = 99009901 = 990^2 + 09901^2.$$

What are the numbers which can be expressed as the sums of two squares? There is not an obvious pattern, and this is a question which we shall explore in a later section of this chapter.

Higher powers of the digits also bear some investigation. To various bases, we can find numbers which are the sums of the cubes of their digits.

Part of a family of such numbers is

$$(251)_7 = 2 \times 7^2 + 5 \times 7 + 1 = 134 = 2^3 + 5^3 + 1^3$$
$$(371)_{10} = 3 \times 10^2 + 7 \times 10 + 1 = 371 = 3^3 + 7^3 + 1^3$$
$$(491)_{13} = 4 \times 13^2 + 9 \times 13 + 1 = 794 = 4^3 + 9^3 + 1^3.$$

These are special cases of the general formula

$$k(3k+1)^2 + (2k+1)(3k+1) + 1 = 9k^3 + 12k^2 + 6k + 2$$
$$= k^3 + (2k+1)^3 + 1^3$$

corresponding to the base $3k + 1$.

The reader may wish to look for other numbers which can be expressed as the sum of the cubes of the digits **(4)**.

We can also consider the mapping that takes any number written in base b to the sums of a fixed power of its digits. Whenever this is done, the sequence obtained eventually ends in a cycle. For example, beginning with any number and continually taking the sum of the squares of the digits to base 10 eventually leads to either a succession of ones or to $\dots, 4, 16, 37, 58, 89, 145, 42, 20, 4, \dots$. For cubes, there are several ending cycles. In particular, each of the base 10 numbers 136 and 244 is the sum of the cubes of the digits of the other; the same holds for 919 and 1459. Each of the numbers 2178 and 6514 is the sum of the fourth powers of the digits of the other. **(5)**

Numbers expressible as the sums of two squares

The list of numbers which are expressible as the sums of two squares may seem haphazard at first, but there is a deeper pattern to be found. The first observation is that the product of any two such numbers is also such a number. This follows from the algebraic identity:

$$(x^2 + y^2)(u^2 + v^2) = (xu - yv)^2 + (xv + yu)^2.$$

Thus, for example, knowing that $5 = 2^2 + 1^2$ and $13 = 3^2 + 2^2$ we can find that

$$65 = (2^2 + 1^2)(3^2 + 2^2) = (2 \times 3 - 1 \times 2)^2 + (2 \times 2 + 1 \times 3)^2$$
$$= 4^2 + 7^2.$$

Taking the sums of the squares of 5 and 13 in different orders will yield the different possible representations of 65 as the sum of two squares. Euler (1707–1783) proved this and other results about numbers expressible as the sums of two squares:

1. A prime number is expressible as the sum of two squares if and only if it is either 2 or leaves a remainder 1 when divided by 4. Thus, 41 can be expressed as the sum of two squares while 43 cannot.
2. A number can be expressed as the sum of two squares if and only if, when it is written as a product of its prime factors, any prime which leaves a remainder 3 when divided by 4, must occur an even number of times.

In 1873, J. W. L. Glaisher **(6)** gave a table that allows one to hone in on those integers expressible as the sum of two positive squares:

1	2	3	4	5	6	7	$\cdots$
-3	-6	-9	-12	-15	-18	-21	$\cdots$
5	10	15	20	25	30	35	$\cdots$
-7	-14	-21	-28	-35	-42	-49	$\cdots$
9	18	27	36	45	54	63	$\cdots$

In the rows, the positive (respectively, negative) multiples of numbers of the form $4k+1$ (respectively, $4k+3$) are written. Given a positive number n, let $+n$ appear P times and $-n$ appear N times; n can be expressed as the sum of two squares if and only if $P > N$.

For primes of the form $4k + 1$, there is a nice algorithm which will allow us to achieve a representation as the sum of two squares. We consider a mapping defined on triples (x, y, z) of numbers by the following equation:

$$U(x,y,z) = \begin{cases} (x-y-z, y, 2y+z), & \text{if } x > y+z; \\ (y, x, -z), & \text{if } x < y+z. \end{cases}$$

(The possibility $x = y + z$ will not occur in the cases we are considering.) The significant property is that the value of $4xy + z^2$ is invariant under U. That is, if $(X, Y, Z) = U(x, y, z)$, then $4XY + Z^2 = 4xy + z^2$. To see how this works, let us take $p = 73$. We write this in the form $p = 4k + 1$; here $k = 18$. Now start with the triple $(k, 1, 1)$ and apply the operator U over and over again until we reach a triple of the form (a, a, b). We will discover that $p = (2a)^2 + b^2$, since, at each stage, the value of $4xy + z^2$ retains its original value of $4k + 1$. For $p = 73$, we apply U to get the chain of triples

$$(18, 1, 1) \to (16, 1, 3) \to (12, 1, 5) \to (6, 1, 7) \to (1, 6, -7)$$

$$\to (2, 6, 5) \to (6, 2, -5) \to (9, 2, -1) \to (8, 2, 3) \to (3, 2, 7)$$

$$\to (2, 3, -7) \to (6, 3, -1) \to (4, 3, 5) \to (3, 4, -5) \to (4, 4, 3).$$

The last triple is of the required form, and we find that $73 = 64 + 9 = (2 \times 4)^2 + 3^2$. The reader may wish to try this on other primes: $5 = 4 \times 1 + 1$;

$13 = 4 \times 3 + 1$; $17 = 4 \times 4 + 1$; $29 = 4 \times 7 + 1$; $37 = 4 \times 9 + 1$; $41 = 4 \times 10 + 1$; $53 = 4 \times 13 + 1$; etc., **(7)**.

Numbers expressible as the sum of three and four squares

We have seen that not every positive whole number can be written as the sum of two squares. Nor can every number be written as the sum of three squares. The ones for which such a representation is impossible are those that are a power of four multiplied by an odd number equal to one less than a multiple of 8. Thus, the numbers 7, 15, 23, 28, 31, 39, 47, 55, 60, etc., all cannot be written as the sum of three squares.

However, it is a remarkable fact, established by J. L. Lagrange that every positive number is the sum of four squares. Because of the identity **(8)**,

$$(a^2 + b^2 + c^2 + d^2)(p^2 + q^2 + r^2 + s^2)$$
$$= (ap - bq - cr - ds)^2 + (aq + bp + cs - dr)^2$$
$$+ (ar + cp + dq - bs)^2 + (as + dp + br - cq)^2,$$

it is enough to prove that every prime number can be written as the sum of four squares, since every positive integer can be written as the product of primes.

It is natural to ask the same question about higher powers. In 1770, Waring conjectured that each number could be expressed as the sum of at most nine cubes and of at most 19 fourth powers. Around the beginning of the twentieth century, David Hilbert proved that, for any positive integer k, there is a number g_k such that every positive integer can be written as the sum of no more than g_k positive kth powers, and there is some integer that requires g_k positive kth powers.

It is now known that each number, in fact, *can* be expressed as the sum of at most nine positive cubes. Only the numbers 23 and 239 cannot be expressed as the sum of fewer than nine cubes. There are fifteen additional numbers, the largest of which is 454, that need eight cubes, and 121 more numbers, topped by 8042, that require seven cubes. If we drop the condition that the cubes in the sum be positive, then it is conjectured that each number can be expressed as the sum of four cubes. For example, $23 = 2^3 + 2^3 + 2^3 + (-1)^3$.

It is still an open question whether $g_4 = 19$. That $g_4 \geq 19$ is clear from the fact that 79 is the sum of 19 fourth powers and canot be expressed as the sum of fewer than 19 fourth powers. It is known that $g_5 = 37$, with 223 requiring 37 positive fifth powers in its sum.

Euler conjectured that, if h_k is the largest integer not exceeding $(3/2)^k$, then $g_k = h_k + 2^k - 2$. This conjecture has been established except for $k = 4$

and possibly finitely many other values of k. If we define G_k to be the smallest number r such that there are infinitely many positive integers that require r positive kth powers in the sum, and every integer from some point on needs no more than r positive kth powers, then, clearly $G_k \leq g_k$. The only values of k for which G_k is known for certain are $k = 2$ and, surprisingly, $k = 4$. For the others, we can only provide bounds; thus, $4 \leq G_3 \leq 7$. The following table gives part of the story (9):

k	g_k	G_k bounds
2	4	4
3	9	$[4, 7]$
4	19	16
5	37	$[6, 23]$
6	73	$[9, 36]$
7	143	$[8, 137]$
8	279	$[32, 163]$
9	548	$[13, 190]$

Digital oddities of powers

Let us pick up the digital theme with which the chapter began. Sticking to base 10, we note that there are exactly two numbers which along with their squares involve each of the nine nonzero digits exactly once:

$$567^2 = 321489 \qquad 854^2 = 729316.$$

Two consecutive powers of 69, $69^2 = 4761$ and $69^3 = 328509$ (10) and two consecutive powers of 18, $18^3 = 5832$ and $18^4 = 104976$ both involve each of the ten digits exactly once.

As a variant on this, the equation $x + y = z^3$, where x, y, and z together involve the nine digits once each, has the solutions $(x, y, z) = (19635, 48, 27)$, $(19638, 45, 27)$, $(19645, 38, 27)$, $(19648, 35, 27)$ (11).

While there are no squares apart from 1, 4, and 9 that make use of only one digit, there are some that use only two digits. Apart from the squares of single digit numbers, 10^n, 2×10^n and 3×10^n, we have

$$38^2 = 1444 \quad 88^2 = 7744 \quad 109^2 = 11881 \quad 173^2 = 29929$$

$$212^2 = 44944 \quad 235^2 = 55225 \quad 3114^2 = 9696996.$$

It seems to be unknown whether there are infinitely many other examples (12).

The example $7744 = 88^2$ has the form $(xxyy)_b = (zz)_b^2$, where b is a base of numeration. Some bases will admit many examples of this type, up to 63 examples, when $b = 9241, 60061, 78541, 87781$, and 92421 (13).

Observe that $9898^2 = 97970404$. The form

$$(mnmn)_b^2 = (ababcdcd)_b$$

is possible for other bases b **(14)**. For example

$$(2323)_4^2 = (20202121)_4$$
$$(3434)_7^2 = (16165252)_7$$
$$(4343)_7^2 = (26264242)_7$$
$$(6565)_7^2 = (64640404)_7$$
$$(4545)_9^2 = (23236767)_9.$$

Victor Thébault asked for what bases b we obtain a numerical equation of the type $(xyx)_b^2 = (zzxxyy)_b$. The only possibility known appears to be $(242)_6^2 = (112244)_6$ **(15)**.

There are at least two ten-digit squares for which the two numbers formed by the first five and the last five digits are consecutive **(16)**:

$$9901^2 = 9802\mathit{9801}$$
$$36365^2 = 13224\mathit{13225}$$
$$63636^2 = 40495\mathit{40496}.$$

We also have the repeated parts

$$4^2 \times 11^2 \times 826446281^2 = 13223140496\mathit{13223140496}.$$

There are 30 squares that use each of the 9 nonzero digits once and 87 that use each of the 10 digits once. For ten digits, the smallest and largest such squares are **(17)**

$$32043^2 = 1026753849 \quad \text{and} \quad 99066^2 = 9814072356.$$

Two for which the square roots are palindromes are

$$97779^2 = 9560732841 \quad \text{and} \quad 84648^2 = 7165283904.$$

The first is one of only four squares whose root has all digits odd; the second is the only square whose root has all digits even. The pythagorean triples $(546, 728, 910)$ and $(534, 712, 890)$ each involve nine distinct digits **(18)**.

Squares themselves can be palindromes:

$$26^2 = 676, \qquad 264^2 = 69696, \qquad 307^2 = 94249,$$
$$836^2 = 698896, \quad \text{and} \quad 798644^2 = 637832238736.$$

Most of them seem to have an odd number of digits; only four with evenly many digits are known **(19)**.

Sometimes the sum and the difference of two numbers, one the digital reverse of the other, are both squares **(20)**:

$$65 - 56 = 3^2 \quad 65 + 56 = 11^2$$
$$621770 - 77126 = 738^2 \quad 621770 + 77126 = 836^2.$$

No cubes have the nine nonzero or all ten digits each occurring exactly once. The largest cube with all digits distinct is **(21)**

$$319^3 = 32461759.$$

This cube shares the same digits as $289^3 = 24137569$.

There is one pair of cubes which together contain all the digits once each: $(21^3, 93^3) = (9261, 804357)$ **(22)**. The four cubes $1^3 = 1$, $2^3 = 8$, $4^3 = 64$ and $59^3 = 205379$ together use each digit once. There are three pairs of cubes that use the nonzero digits once each; one such is $(5^3, 76^3) = (125, 438976)$.

The last ten digits of each of the cubes

$$2326^3 = 12584301976 \quad \text{and} \quad 7616^3 = 441754320896$$

have one occurrence of each of the ten digits. The last nine digits of 3024^3, 5032^3, and 9463^3 are distinct and nonzero **(23)**. There are 138 cubes that contain each of the ten digits exactly twice, including

$$2158479^3 = 10056421854778936239$$
$$4631793^3 = 99368200745116834257.$$

Others contain each of the nine nonzero digits twice:

$$496536^3 = 122419957378438656$$
$$657756^3 = 284573499861537216.$$

Note that the cube root of the second example is a palindrome **(24)**.

Each of the following pairs of powers possess all ten digits: $(188682^2, 188683^2)$, $(28^{11}, 29^{12})$, as do each of the triples $(82^9, 83^9, 84^9)$ **(25)**.

The sum of squares

$$1^2 + 2^2 + \cdots + 2187^2 = 3489176250$$

contains each digit exactly once, while the sum

$$1^2 + 2^2 + \cdots + 181^2 = 1992991$$

is a palindrome **(26)**.

A nice interplay of digits, summation and cubing is given by **(27)**:

$$1212\mathit{1388}2349 = (1212 + 1388 + 2349)^3$$
$$1287\mathit{1113}2649 = (1287 + 1113 + 2649)^3$$
$$1623\mathit{2457}1375 = (1623 + 2457 + 1375)^3$$
$$1713\mathit{2377}1464 = (1713 + 2377 + 1464)^3$$
$$3689\mathit{1035}2448 = (3689 + 1035 + 2448)^3.$$

Observe also that **(28)**

$$54199^3 = 15921\mathit{1275}242599 \qquad 15921 + 12752 + 42599 = 71272$$
$$71272^3 = 362040\mathit{234}715648 \qquad 36204 + 02347 + 15648 = 54199.$$

Automorphic numbers of degree n are those whose nth powers end in themselves. Thus 749 is automorphic of degree 3 since $749^3 = 420189\mathit{749}$. Other numbers that are automorphic of degree 3 are 24, 25, 49, 51, 75, 76, 99, 251, 501, 751, 49999, 50001. In particular, $49999^3 = 124992500\mathit{149999}$ and $50001^3 = 125007500\mathit{150001}$. The numbers 32, 43, 51, and 93 are automorphic of degree 5 **(29)**.

There are several powers whose digits are permutations of other powers. Some small examples are:

$$2^8 = 4^4 = 256, \qquad 5^4 = 625,$$
$$5^3 = 125, \quad 2^9 = 8^3 = 512,$$
$$2^{10} = 1024, \quad 7^4 = 2401,$$
$$3^5 = 243, \quad 18^2 = 324,$$
$$35^3 = 42875, \quad 38^3 = 54872,$$
$$13^2 = 169, \quad 14^2 = 196, \quad 31^2 = 961,$$
$$6^4 = 1296, \quad 54^2 = 2916, \quad 96^2 = 9216, \quad 21^3 = 9261.$$

The powers 2^{14}, 178^2, 191^2, 14^4, and 209^2 are all permutations of the digits $\{1, 3, 4, 6, 8\}$ **(30)**, and the powers 2^{20}, 1028^2, 1042^2, 2396^2, 7^8, 2599^2, and 2801^2 are permutations of $\{0, 1, 4, 5, 6, 7, 8\}$.

Herman Nijon **(31)** gives example of powers the sums of whose digits is equal to the base. For example, $512 = (5 + 1 + 2)^3$, $2401 = (2 + 4 + 0 + 1)^4$, and $(1 + 7 + 2 + 1 + 0 + 3 + 6 + 8)^5 = 17210368$. A table of powers and bases exhibiting this property is:

Exponent	Base
2	9
3	$8, 17, 18, 26, 27$
4	$7, 22, 25, 28, 36$
5	$28, 35, 36, 46$
6	$18, 45, 54, 64$
7	$18, 27, 31, 34, 43, 53, 58, 68$
8	$46, 54, 63$
9	$54, 71, 81$
10	$82, 85, 94, 97, 106, 117$
12	108

C. W. Trigg **(32)** draws attention to the oddities

$$3^2 = 9 \quad 3^3 = 27 \quad 3^{2 \times 3} = 729$$

$$3^3 = 27 \quad 3^4 = 81 \quad 3^{3+4} = 2187.$$

Another striking equation involving powers is $2^5 9^2 = 2592$ **(33)**.

It is left as an exercise for the reader to discover the reason behind these patterns:

$$56^2 - 45^2 = 1111; \quad 556^2 - 445^2 = 111111; \quad 5556^2 - 4445^2 = 11111111; \ldots$$

and

$$4^2 = 16; \quad 34^2 = 1156; \quad 334^2 = 111556; \quad 3334^2 = 1111556; \ldots.$$

We conclude with the observation **(34)**

$$336 = (3 + 3 + 6) + (3^2 + 3^2 + 6^2) + (3^3 + 3^3 + 6^3).$$

Exercises on the Notes

1. *High school algebra; straightforward search.* (a) Let n be a m-digit number that is equal to the sum of the cubes of its digits (to base 10). Prove that $10^{m-1} \leq n \leq 729m$ and deduce that n cannot have more than 4 digits.

(b) Show that, if n has four digits, then the first digit cannot exceed 2. Determine all four-digit numbers equal to the sum of the cubes of their digits.

(c) Determine all numbers not exceeding 100 equal to the sum of the cubes of their digits.

(d) Let n be a three-digit number equal to the sum of the cubes of its digits, i.e., $n = 100a + 10b + c = a^3 + b^3 + c^3$ for $0 \leq a, b, c \leq 9, a \neq 0$.

Rewrite this equation in the form

$$a(10 - a)(10 + a) = b(b^2 - 10) + c(c + 1)(c - 1)$$

and use this to help you search for examples. (Examine the possible values of each term in the equation.)

2. *High school algebra; straightforward search.* Let b be a positive integer exceeding 1, and let

$$n = \cdots + ub^4 + vb^3 + wb^2 + xb + y = \cdots + u^2 + v^2 + w^2 + x^2 + y^2$$

be a number equal to the sum of the squares of its digits to base b.

(a) Suppose that such a number has k digits. Argue that it is at least b^{k-1} and at most $k(b - 1)^2$, and thence deduce that

$$b^{k-1} \leq k(b - 1)^2.$$

(b) Verify that $b^k/(k+1)$ divided by b^{k-1}/k exceeds 1 when $k \geq 2$ and deduce that, if $b^{k-1} < k(b - 1)^2$ fails for any value of k, then it fails for all greater values of k.

(c) Argue that, if m satisfies $b^{m-1} \geq m(b-1)^2$, then any number equal to the sum of the squares of its digits must have fewer than m digits. In particular, show that, if $b = 10$, then the number n cannot have more than three digits.

(d) Prove that 1 is the only number equal to the sum of the squares of its digits to base 10.

3. *High school algebra and number theory; moderate.* Suppose that $xb+y = x^2 + y^2$ is a two-digit number to base b expressible as the sum of the squares of its digits.

(a) Verify that $xb + y = x^2 + y^2$ is equivalent to

$$b^2 + 1 = (b - 2x)^2 + (2y - 1)^2.$$

(b) Deduce from (a) that if $b^2 + 1 = u^2 + v^2$ for integers u and v for which b and u have the same parity and v is odd, then it is possible to find values of x and y satisfying the equation in (a).

(c) Use (b) and b equal to a power of 10 to construct further examples of numbers of the type 5882353.

6. *High school number theory; moderate.* (a) Observe that n will appear in row r if and only if r is odd and $2r - 1$ divides evenly into n.

(b) Observe that $-n$ will appear in row r if and only if r is even and $2r - 1$ divides evenly into n.

(c) Let $n = 2^a b$ where a is a nonnegative integer and b is odd. Use Euler's result to argue that n is the sum of two squares if and only if b is. Deduce that we need check Glaisher's result only for odd values of n.

(d) Show that, if all the primes dividing the positive number n leave remainder 1 when divided by 4, then n appears in the table and $-n$ never does.

(e) Let $n = q^a b$ where q is a prime leaving remainder 3 upon division by 4, a is a nonnegative integer and b is an odd integer. For each integer r with $0 \le r \le a$, let S_r be the set of divisors of n of the form $q^r s$ where s and q are coprime. Verify that for $0 \le r \le a - 1$, there is a one-one correspondence between S_r and S_{r+1} given by $d \leftrightarrow qd$ such that each divisor of the form $4k + 1$ corresponds to a divisor of the form $4k + 3$. Deduce that, if a is odd, $P = N$, while if a is even, $P > N$.

(f) Apply Euler's criterion to establish Glaisher's assertion.

7. *High school mathematics; moderate.* In this problem $(X, Y, Z) = U(x, y, z)$. We will refer to the sequence of vectors obtained by applying U repeatedly to $(n, 1, 1)$ as the orbit of $(n, 1, 1)$.

(a) Verify that $4XY + Z^2 = 4xy + z^2$.

(b) Show that, if $x < y + z$, the $X > Y + Z$, so that the second option is not taken twice in a row in the calculation of the orbit of $(n, 1, 1)$.

(c) What happens to the orbit of $(n, 1, 1)$ when $4n + 1$ is a square?

(d) What happens to the orbit of $(n, 1, 1)$ when $4n + 1$ is composite? Do we ever get a representation of $4n + 1$ as the sum of two squares?

(e) When $4n + 1$ is prime, show that $x \ne y + z$ and $xyz \ne 0$ for each (x, y, z) in the orbit of $(n, 1, 1)$.

(f) When $4n + 1$ is prime, show that $0 < x \le n$, $0 < y \le n$, $-2\sqrt{n} < z < 2\sqrt{n}$ for (x, y, z) in the orbit of $(n, 1, 1)$. Deduce from this that not all the elements of the orbit are distinct and so the orbit is ultimately periodic (i.e., consists of the same entries repeated in cyclic fashion).

(g) For the case that $4n + 1$ is prime, continue the orbit until a vector (x, y, z) appears for the second time. Examine the sequence obtained and make some conjectures. Try to prove them.

13. *High school theory of the quadratic and basic number theory; difficult.* Suppose that $(xxyy)_b = (yy)_b^2$. Prove that y and $x^2 + 4(x - 1)^2$ are both squares. Determine an example for which x is also square.

19. Prove that the product of any two-digit number with distinct digits and its reversal, obtained by writing the digits in reverse order, is never a square.

Notes

1. For each positive integer k, there are only finitely many numbers equal to the sum of the kth power of their digits. Such a number with n digits must be at least 10^{n-1} but not exceed $n \times 9^k$. Hence

$$\frac{10^{n-1}}{n} \le 9^k$$

whence

$$n\left[1 - \frac{1 + \log_{10} n}{n}\right] = n - (1 + \log_{10} n) \le k \log_{10} 9 < 0.96k.$$

For $n \ge 72$, $(1 + \log_{10} n)/n < 0.4$ and so $n < k$. Thus, the number of digits of the number cannot exceed the maximum of $k - 1$ and 72. A finer analysis will give much better estimates.

For example, if $k = 4$, we have $10^{n-1}/n \le 9^4 = 6561$, whence $n \le 5$. For $n = 5$, the sum of the fourth powers of the digits is at most $5 \times 6561 = 32805$, so the first digit cannot exceed 3. This imposes a stronger upper bound of $3^4 + 4 \times 9^4 = 26325$ for the number. For $n = 6$, similar analysis shows that the number cannot exceed 298370. A list of socalled "powerful" numbers appears in *Mathematical Gazette* 52 (1968), 383.

2. If a k-digit number to base $b \ge 2$ is the sum of the squares of its digits, then $b^{k-1} \le k(b - 1)^2$, or $b^{k-1}/k \le (b - 1)^2$. Let $k \ge 2$. Since $(b^k/(k+1)) \div (b^{k-1}/k) = bk/(k+1) \ge 2 \cdot 2/3 > 1$, it is clear that b^{k-1}/k is an increasing function of k. Thus, the number of digits is bounded by the smallest value of m for which $b^m \ge (m + 1)(b - 1)^2$. In particular, when $b = 10$, the number cannot have more than three digits.

It is easy to dispose of the three-digit case. The number cannot exceed $3 \times 9^2 = 243$, and so cannot exceed $2^2 + 2 \times 9^2 = 166$. Since $1^2 + 2 \times 7^2 = 99$, it follows that the last digit must be either 8 or 9. Checking the possibilities yields no examples. The two-digit case is more tedious, but can be handled similarly.

3. See *Mathematics Magazine* 57 (1984), 236–237 and *Journal of Recreational Mathematics* 2 (1969), 109–110; 3 (1970), 186–187. It is not hard to construct more numbers of this type. Suppose that b is the base of numeration, where now we take b to be a power of ten. Then we are looking for numbers x and y for which

$$xb + y = x^2 + y^2$$

or equivalently

$$b^2 + 1 = (b - 2x)^2 + (2y - 1)^2.$$

The number $b^2 + 1$ can be expressed as the sum of two squares in at least one way. Let $b^2 + 1 = u^2 + v^2$ be any such representation for which b and u have the same parity and v is odd. We solve the equations

$$b - 2x = \pm u \qquad 2y - 1 = \pm v$$

for x and y. We can have

$$(x, y) = \left(\tfrac{1}{2}(b - u), \tfrac{1}{2}(v + 1)\right) \quad \text{or} \quad (x, y) = \left(\tfrac{1}{2}(b + u), \tfrac{1}{2}(v + 1)\right).$$

For example, starting with $b = 10^2$ and $10^4 + 1 = 76^2 + 65^2$, we are led to the possibilities $(x, y) = (12, 33)$ and $(x, y) = (88, 33)$ for the numbers $1233 = 12^2 + 33^2$ and $8833 = 88^2 + 33^2$.

4. Problem 4423 in *American Mathematical Monthly* 58 (1951), 42; 59 (1952), 415–416 asked for which bases of numeration there exist pairs of consecutive three-digit numbers each equal to the sum of the cubes of its digits. The solution uses Pell's equation to find infinitely many such possibilities; in each case, the units digits of the two numbers are 0 and 1. For example $(660)_8 = 432$ and $(661)_8 = 433$ are each the sum of the cubes of their digits. A general formula for the smaller of such a pair is

$$(k + 1)(3k + 1)\left[(3k + 1)^2\right]^2 + (2k + 1)(3k + 1)\left[(3k + 1)^2\right]$$
$$= (k + 1)^3(3k + 1)^3 + (2k + 1)^3(3k + 1)^3,$$

where the base is $(3k + 1)^2$.

5. David E. Kullman's paper [Sums of powers of digits, *Journal of Recreational Mathematics* 14 (1981–1982), 4–10] has a good collection of references and a list of cycles of numbers, each of which is the sum of the kth powers of the digits of its predecessor for small values of k.

Lionel E. Deimel, Jr. and Michael T. Jones [Finding pluperfect digital invariants: techniques, results and observations, *Journal of Recreational Mathematics* 14 (1981–1982), 87–107, 284] define a number to be "pluperfect digital invariant" if it is the sum of the kth power of its digits to some base b, where k is the number of digits it has. An example is the 14-digit number 28116440335967 which is the sum of the 14th powers of its digits. The paper contains a table of examples for bases from 2 to 10 that run to several pages.

6. See Volume II of Dickson's *History*, page 243.

7. Call the determination of $U(x, y, z)$ by $(x - y - z, y, z + 2y)$ when $x > y + z$ the *first option*, and by $(y, x, -z)$ when $x < y + z$ the *second option*. Let $U(x, y, z) = (X, Y, Z)$. Given a prime $p = 4n + 1$, we wish to show that U applied repeatedly, beginning with $(n, 1, 1)$ will yield the form (r, r, s), so that $p = 4r^2 + s^2$, the sum of two squares.

To illustrate the phenomenon under study, consider the example $n = 15$. The successive iterates of $(15, 1, 1)$ are

$$(15, 1, 1) \to (13, 1, 3) \to (9, 1, 5) \to (3, 1, 7) \to (1, 3, -7) \to (5, 3, -1)$$

$$\to (3, 3, 5) \to (3, 3, -5) \to (5, 3, 1) \to (1, 3, 7) \to (3, 1, -7)$$

$$\to (9, 1, -5) \to (13, 1, -3) \to (15, 1, -1) \to (15, 1, 1) \to \cdots.$$

Note that the sequence ultimately returns to its starting point and that it is "almost" symmetric in that, up to the sign of the third coordinate, it reads the same backwards.

Return to the general case. Define $\alpha_1 = (n, 1, 1)$, $\alpha_k = U^{k-1}(n, 1, 1)$ for $k \geq 2$. If $\alpha_k = (x, y, z)$, let $\alpha_{-k} = (x, y, -z)$ for each positive integer k. We establish the desired result through a sequence of propositions.

(1) U leaves the form $4xy + z^2$ invariant, i.e., $4XY + Z^2 = 4xy + z^2$.

(2) If $x < y + z$, then $X > Y + Z$, so that the second option does not occur for two consecutive iterates of U.

(3) $x = y + z$ can never occur in the orbit $\{U^k(n, 1, 1) : k = 0, 1, 2, \ldots\}$ of $(n, 1, 1)$. Thus, the value of U is always defined along the orbit. [Otherwise, $p = 4xy + z^2 = (2y + z)^2$.]

(4) $xyz \neq 0$ for each (x, y, z) in the orbit of $(n, 1, 1)$ [since p is neither even nor square].

(5) $0 < x \leq n$, $0 < y \leq n$, $-2\sqrt{n} < z < 2\sqrt{n}$ for each (x, y, z) in the orbit of $(n, 1, 1)$. Hence, the orbit is bounded and ultimately periodic, i.e., it must at some stage return to a previous element. [Observe that $\max(4xy, z^2) < 4xy + z^2 = 4n + 1$.]

(6) $U(\alpha_{-1}) = \alpha_1$.

(7) For each $k \geq 1$, $U(\alpha_{-(k+1)}) = \alpha_{-k}$.

Proof of (7) in case $\alpha_k = (x, y, z)$ with $x > y + z$. Then

$$\alpha_{k+1} = (x - y - z, y, z + 2y),$$

so $\alpha_{-(k+1)} = (x-y-z, y, -z-2y)$. Since $(x-y-z)-y-(-z-2y) = x > 0$, it follows that $U(\alpha_{-(k+1)}) = (x, y, -z - 2y + 2y) = (x, y, -z) = \alpha_{-k}$.

Proof of (7) in case $\alpha_k = (x, y, z)$ with $x < y + z$. Then $\alpha_{k+1} = (y, x, -z)$ and $\alpha_{-(k+1)} = (y, x, z)$. We need to show that $y < x + z$, from which it follows that $U(\alpha_{-(k+1)}) = (x, y, -z) = \alpha_{-k}$. Since the second option cannot occur twice in a row,

$$\alpha_{k-1} = (u, v, w) \quad \text{where } u > v + w.$$

We have that $x = u - v - w$; $y = v$; $z = w + 2v$, whence $u = x - y + z$; $v = y$; $w = z - 2y$. Since $u > 0$, we have that $x - y + z > 0$, or $x + z > y$, as desired.

Having disposed of these preliminary results, we are in a position to obtain the representation of p as the sum of two squares. Suppose that α_m is the first element in the orbit of $(n, 1, 1)$ to appear twice. Then $\alpha_m = U(\alpha_{m-1}) = U(\alpha_s)$ for some $s \geq m$. From the above results, we can infer the following orbit diagram for the mapping U.

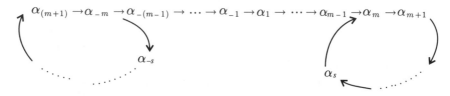

FIGURE 6.1

If $m > 1$ we are led to the impossibility that α_{-m} has two images under U. The only way out is that $m = 1$ and $\alpha_s = \alpha_{-1}, \alpha_{s-1} = \alpha_{-2}, \ldots,$ $\alpha_{s-t} = \alpha_{-t-1}, \ldots, \alpha_1 = \alpha_{-s}$.

Suppose that s is odd. Then, taking $t = \frac{1}{2}(s - 1)$ yields $\alpha_{(s+1)/2} = \alpha_{-(s+1)/2}$, so that the third coordinate of $\alpha_{(s+1)/2}$ vanishes, an impossibility.

Thus, s is even. Taking $t = \frac{s}{2}$ yields $\alpha_{s/2} = \alpha_{-s/2-1}$. If $\alpha_{s/2} = (x, y, z)$ with $x > y + z$, then $\alpha_{s/2+1} = (x - y - z, y, z + 2y)$ and $\alpha_{-s/2-1} = (x-y-z, y, -z-2y) = \alpha_{s/2}$. Hence $y + z = 0$ and $p = 4xy + z^2 = y(4x+y)$. Since p is prime, $y = 1$ and $x = n$. Thus $\alpha_{s/2} = (n, 1, -1) = \alpha_{-1} = \alpha_s$, an impossibility.

Therefore, $\alpha_{\frac{s}{2}} = (x, y, z)$ with $x < y + z$, so $\alpha_{\frac{s}{2}+1} = (y, x, -z)$, whence $(y, x, z) = (x, y, z)$ and $x = y$, as desired.

The article *Three paths to Mt. Fermat-Euler* by Vladimir Tikhomirov [*Quantum* 4 (no. 5), (May/June, 1994), 5–7] discusses a similar technique based on the transformation of Zagier:

$$B(x, y, z) = \begin{cases} (x + 2z, z, y - x - z), & \text{if } x < y - z \\ (2y - x, y, x - y + z), & \text{if } y - z \leq x < 2y \\ (x - 2y, x - y + z, y), & \text{if } 2y \leq x \end{cases}$$

which leaves invariant $x^2 + 4yz$. The same paper provides two other more standard arguments that every prime exceeding a multiple of 4 by 1 is the sum of two squares. See also the note by D. Zagier [*American Mathematical Monthly* 97 (1990), 144].

In Note 2733 [*Mathematical Gazette* 41 (1957), 288–289], A. Sutcliffe established that an integer can be expressed as the sum of two squares if and only if it is of the form $2^r(4s+1)$, where s is the sum of two triangular numbers. Another proof by M. D. Hirshhorn appears in *American Mathematical Monthly* 92 (1985), 579–580.

8. The identity $(x^2+y^2)(u^2+v^2) = (xu-yv)^2+(xv+yu)^2$ is an expression of the relation $|zw| = |z||w|$ where $z = x + yi$ and $w = u + vi$ are complex numbers. In a similar vein, we can assign significance to

$$(a^2 + b^2 + c^2 + d^2)(p^2 + q^2 + r^2 + s^2)$$
$$= (ap - bq - cr - ds)^2 + (aq + bp + cs - dr)^2$$
$$+ (ar + cp + dq - bs)^2 + (as + dp + br - cq)^2.$$

We define "hypercomplex" numbers called quaternions. This is a family H of expressions of the form $a + bi + cj + dk$ where a, b, c, d are real numbers and addition, multiplication, and subtraction are defined in the usual way, subject to i, j, k being abstract entities satisfying the relations $i^2 = j^2 = k^2 = -1$, $ij = k$, $jk = i$, $ki = j$. All the laws of arithmetic hold for H except for commutativity of multiplication. Indeed, we find that $ji = -k$, $ik = -j$ and $kj = -i$. As for complex numbers, we can define an absolute value

$$|a + bi + cj + dk| = \sqrt{a^2 + b^2 + c^2 + d^2}$$

and we find that $|zw| = |z||w|$ for $z, w \in H$.

For a treatment of the four square theorem, consult the articles by J. D. Dixon [*American Mathematical Monthly* 71 (1964), 286–288] and by André Weil [*American Mathematical Monthly* 91 (1984), 537–542]. Results on numbers expressible as sums of squares in various ways appear in *American Mathematical Monthly* 40 (1933), 10–18.

9. It is not hard to show that each natural number can be written as the sum of 53 fourth powers, possibly including 0, given that each number is expressible as the sum of four squares. Begin with the identity, due to Liouville (1859):

$$6(x_1^2 + x_2^2 + x_3^2 + x_4^2)^2 = \sum_{1 \leq i < j \leq 4} (x_i + x_j)^4 + \sum_{1 \leq i < j \leq 4} (x_i - x_j)^4$$

from which we deduce that each number of the form $6k^2$ can be written as the sum of 12 fourth powers. For any number m, $6m$ can be written as the sum of four numbers of the form $6k^2$ and hence as the sum of 48 fourth powers. Any positive integer whatsoever has the form $6m + i$ where $0 \le i \le 5$, which can be expressed as the sum of 48 fourth powers plus i ones and $5 - i$ zeros for a total of 53 fourth powers.

Hardy and Littlewood showed that each number can be written as the sum of no more than $(n - 2)2^{n-1} + 5$ nth powers. For $n = 2, 3, 4$, this upper bound takes the respective values 5, 9, 21.

See the survey papers by P. M. Batchelder, W. J. Ellison and C. Small in, respectively, *American Mathematical Monthly* 43 (1936), 21–27; 78 (1971), 10–36 and *Mathematics Magazine* 50 (1977), 12–16.

In the case of cubes, we can, as an alternative, allow negative as well as nonnegative cubes in the representation. In this case, it appears that each integer can be represented as the sum of four cubes. For a history and discussion of this problem, consult the article by D. A. Klarner [*American Mathematical Monthly* 74 (1967), 531–537].

10.   See Problem E116 in *American Mathematical Monthly* 41 (1934), 517; 42 (1935), 175–176.

11.   See Problem 4114 in *American Mathematical Monthly* 51 (1944), 168; 52 (1945), 346–347.

12.   This question is posed on page 146 of Richard Guy's *Unsolved problems in number theory*. See also *Journal of Recreational Mathematics* 25 (1993), 1–5.

13.   See Problems E310 and 4184 in *American Mathematical Monthly* 45 (1936), 47; 46 (1939), 110–111; 52 (1945), 582; 54 (1947), 237. We have to solve

$$xb^3 + xb^2 + yb + y = (zb + z)^2$$

which reduces to

$$x(b - 1) + \frac{x + y}{b + 1} = z^2.$$

Since $(x + y)/(b + 1)$ must be an integer and $x + y \le 2(b - 1) < 2b$, it follows that $x + y = b + 1$. Hence, we must have $z^2 \equiv 1 \pmod{b - 1}$.

Specializing to the case $z = y$, and using $y = b + 1 - x$, we obtain

$$(b + 1 - x)^2 = x(b - 1) + 1 \quad \text{or} \quad b^2 + (2 - 3x)b + (x^2 - x) = 0.$$

As a quadratic in b, this equation has an integer root and so its discriminant $(2-3x)^2 - 4(x^2 - x) = x^2 + 4(x-1)^2$ must be a square. Similarly, setting $x = b+1-y$ yields $b^2 - yb + (y-y^2) = 0$, whose discriminant $y^2 - 4(y-y^2) = y(5y-4)$ must be a square. Since the greatest common divisor of y and $5y-4$ must divide 4, we conclude that either both y and $5y-4$ are squares or $y = 2r^2$ and $5y - 4 = 2s^2$ for some integers r and s. In the latter case, $5r^2 - 2 = s^2$, which is not possible modulo 8. An example is $(x, y; b) = (16, 25; 40)$.

14. See Problem 4157 in *American Mathematical Monthly* 52 (1945), 220; 53 (1946), 473 and Problem 4144 in *American Mathematical Monthly* 51 (1944), 593; 53 (1946), 284–285.

15. See Problem 4227 in *American Mathematical Monthly* 53 (1946), 594; 55 (1948), 507–508.

16. See Problem E582 in *American Mathematical Monthly* 50 (1943), 454; 51 (1944), 231–232. In *Mathematical Gazette* 30 (1946), 19–21, T. R. Dawson gives many examples of "ornamental" square and triangular numbers.

17. For related material, see Problems E389, E414 and 3296 in *American Mathematical Monthly* 46 (1939), 512, 515; 47 (1940), 175, 321, 711; 48 (1941), 281. A list of all squares with nine nonzero or all ten distinct digits is given in *Journal of Recreational Mathematics* 1 (1968), 62–64. This article is followed up in *Journal of Recreational Mathematics* 18 (1985–1986), 95–100. The last two *Monthly* problems give a number of squares which, when reduced by 1, give numbers with all ten digits each occurring once. See also problems 1106, 1108, and 1127 in *Journal of Recreational Mathematics* 14 (1981–1982), 222, 299; 15 (1982–1983), 235–236, 308–309.

18. *Mathematics Magazine* 40 (1967), 221; 41 (1968), 96.

19. See the following papers, all from the *Journal of Recreational Mathematics*:

Charles Aschbacher, More on palindromic squares, JRM 22 (1990), 133–135.

Michael Keith, Classification and enumeration of palindromic squares, JRM 22 (1990), 124–132.

J. Keith and R. Barnett, Tables of square palindromes in bases 2 and 10, JRM 23 (1991), 13–18.

Problem E1243 [*American Mathematical Monthly* 63 (1956), 724; 64 (1957), 434–435] requires it to be shown that the product of any two-digit number with distinct digits and its reversal is never a square. To see this, let m and n be the two numbers, each the reversal of the other. By casting out nines, we find that $m - n$ is a multiple of 9 with the result that the greatest common divisor of m and n is 1, 3, or 9. If it is 1 or 9, then, since mn is square, each of m and n must be square. But the reversal of no two-digit square is square. If the greatest common divisor is 3, then $m = 3u^2$, $n = 3v^2$; it is readily checked that this is not possible.

However, we have that $(144)(441) = 252^2$, $(169)(961) = 403^2$ and $(1089)(9801) = 3267^2$. It is conjectured that, whenever an integer and its reversal are unequal, then their product is a square if and only if both are squares themselves.

20.   The question of finding other examples is posed among others on reversals in *American Mathematical Monthly* 96 (1989), 425–428.

21.   Consult the articles by Charles W. Trigg:

Cubes with distinct digits, *Journal of Recreational Mathematics* 15 (1982 –1983), 285–288

9 and 10 distinct digits in square integers, *Journal of Recreational Mathematics* 18 (1985–1986), 95–100

See also problems 1097, 1194, 1213 and 1369 in the same journal (14 (1981–1982), 220; 15 (1982–1983), 143, 222, 230; 16 (1983–1984), 150, 229; 17 (1984–1985), 56; 18 (1985–1986), 66).

Trigg also lists powers of primes that have no repeated digits in *Journal of Recreational Mathematics* 15 (1982–1983), 14–16.

22.   See Problem E377 in *American Mathematical Monthly* 46 (1939), 236; 47 (1940), 178–179.

23.   See Note 1823 in *Mathematical Gazette* 29 (1945), 80.

24.   See problems 1126 and 1245 in the *Journal of Recreational Mathematics* 14 (1981–1982), 298; 15 (1982–1983), 301, 308; 16 (1983–1984), 312–313.

25.   See Problem 1369 in the *Journal of Recreational Mathematics* 17 (1984 –1985), 56; 18 (1985–1986), 66.

26.   See Problem 1416 in the *Journal of Recreational Mathematics* 17 (1984 –1985), 216; 18 (1985–1986), 303.

27. See Problem 4164 in *American Mathematical Monthly* 52 (1945), 346; 53 (1946), 539–540.

28. See Problem 1377 in the *Journal of Recreational Mathematics* 17 (1984 –1985), 58–59; 18 (1985–1986), 72–74.

29. *Journal of Recreational Mathematics* 25 (1993), 1–5.

30. These are given by C. W. Trigg in Problem 1501, *Journal of Recreational Mathematics* 18 (1985–1986), 239; 19 (1987), 237–238. See also Problem E377 in *American Mathematical Monthly* 47 (1940), 179, item V for cubic examples.

31. See *Journal of Recreational Mathematics* 12 (1979–1980), 293–294.

32. Problem 1493, *Journal of Recreational Mathematics* 18 (1985–1986), 237; 19 (1987), 156.

33. In Q585, *Mathematics Magazine* 47 (1974), 48, 60, it is claimed that $x^y z^x = (xyzx)_b$ occurs only when $(x, y, z; b) = (2, 5, 9; 10)$.

34. Michael Keith, Power-sum numbers *Journal of Recreational Mathematics* 18 (1985–1986), 275–278.

Additional Exercises

1. For what bases of numeration are there perfect squares of the form $(xy)_b$, $(yy)_b$, and $(zw)_b$, where x, y, z, and w are consecutive integers?

2. For what bases of numeration can a four-digit number $(xxyz)_b$ be the square of a two-digit number $(mn)_b$ where $z = y + 1$ and $n = m + 1$?

3. What is the smallest base b for which there is a perfect cube of the form $(xyxy)_b$?

4. Determine a square of at least four digits (in base 10), of which the last is not zero, so that the last three digits, when reversed, form a cube.

5. For which bases b of numeration does there exist at least one four-digit square of the form $(xxyy)_b$ for which the sum of the digits of the square root is $x + y$.

6. Note that $4 = 2^2$, $9 = 3^2$, and $49 = 7^2$. Determine other examples of bases b for which there exists a two-digit number $(xy)_b$ which is square along with each of its digits.

7. If the digits of each of the squares $36 = 6^2$ and $3136 = 56^2$ are all reduced by 1, we get the squares $25 = 5^2$ and $2025 = 45^2$. Show how to construct other squares with the property that reducing each digit by 1 will yield squares.

8. A positive integer is said to be "sorted" if the digits in its decimal representation are nondecreasing from left to right. For example, 3336667 and its square 11133346668889 are both sorted. Prove that, if x is an integer whose base 10 representation consists of an arbitrary number of 3s followed by an arbitrary number of 6s followed by a single 7, then x^2 is sorted.

9. Let S_n be the sum of the digits of 2^n, written to any base. Prove or disprove: $S_{n+1} = S_n$ for some positive integer n.

10. Prove that the numbers 49, 4489, 444889, ..., obtained by inserting 48 into the middle of the preceding numbers are all squares.

11. Show that every rational number r can be written as the sum of exactly four nonzero rational cubes.

12. Suppose that n and m are integers with $n \geq 3$ and $m \geq 2$. Show in a simple way how n^m can be expressed as the sum of n squares.

13. Find the smallest positive integer n which is not a power of 10 such that (in base 10) the initial digits of n^4 are those of n in the same order.

14. In the number system with base 12, every square number ends in a digit which itself is a square. Find all bases for which this is true.

15. In the following message, each letter stands for a particular digit, different letters corresponding to different digits. Determine all solutions for which each of the four numbers is square:

$$MERRY \quad XMAS \quad TO \quad ALL.$$

16. (a) Determine a base for which there is a square consisting of four identical digits, each digit itself being a square.

(b) Provide examples in various bases of squares with four identical digits, each digit not a square.

17. (a) Find perfect squares of seven and eight digits with all digits even and positive.

(b) Show that there are infinitely many perfect squares with all digits even and positive.

(c) Determine all perfect squares with all digits even and distinct.

(d) Show that, except for 1 and 9, there are no perfect squares all of whose digits are odd.

18. (a) Show that the numbers 729, 71289, 7112889, 711128889, etc., are all perfect squares.

(b) Show that the numbers

$$7744, \ 97970404, \ 997997004004, \ 9997999700040004, \text{etc.},$$

are all perfect squares.

(c) Generalize (b) to bases $b \geq 5$ other than 10.

19. J. V. Chaudhari and M. N. Deshpande, in a letter in *Mathematics Teacher* 89 (1996), 83, draw attention to the novel fact that

$$956^2 = 913,936 \qquad 913 + 936 = 1849 = 43^2$$
$$957^2 = 915,849 \qquad 915 + 849 = 1764 = 42^2$$

and so on through consecutive squares down to

$$968^2 = 937,024 \qquad 937 + 024 = 961 = 31^2.$$

They ask for a proof of this and wonder whether there are other strings of squares exhibiting a similar phenomenon.

Solutions to Additional Exercises

1. Let a be any natural number and let $b = 4a + 3$. Then $(x, y, z, w) = (a, a + 1, a + 2, a + 3)$ works since

$$a(4a + 3) + (a + 1) = (2a + 1)^2$$
$$(a + 1)(4a + 3) + (a + 1) = (2a + 2)^2$$
$$(a + 2)(4a + 3) + (a + 3) = (2a + 3)^2.$$

[Problem E572, *American Mathematical Monthly* 50 (1943), 326; 51 (1944), 92–93.]

2. See Problem E652 in *American Mathematical Monthly* 52 (1945), 42.

3. We require that $xb^3 + yb^2 + xb + y = (b^2 + 1)(xb + y)$ be a cube. Since $xb + y \leq (b - 1)b + (b - 1) = b^2 - 1 < b^2 + 1$, $b^2 + 1$ must be divisible by the square of an odd prime. (If $b^2 + 1$ is square free, then the square of every prime dividing $b^2 + 1$ must divide $xb + y$; this makes $xb + y \geq b^2 + 1$ and yields a contradiction.) The smallest numbers divisible by odd prime squares are 9, 18, 25, 27, 36, 45, 50, 54, 63, ...; 50 is the only one of the form $b^2 + 1$ (with $b = 7$). Since $b^2 + 1 = 50 = 2 \times 5^2$, we choose x and y so that $7x + y = 2^2 \times 5 = 20$; $(x, y) = (2, 6)$ works. We find that $(2626)_7 = (13)_7^3$.

4. Since squares cannot end in 2, 3, 7, or 15, the last three digits can only be 521. Some possibilities are $39^2 = 1521$, $211^2 = 44521$, $289^2 = 83521$. In *American Mathematical Monthly* 55 (1948), 20–22, P. A. Pizá shows that the numbers of the form $(250k \pm 39)^2$ are precisely the desired squares.

5. See Problem 4543 in *American Mathematical Monthly* 60 (1953), 423; 61 (1954), 646–647. This problem does not appear to be completely solved. The equation to be considered has the form

$$(xb^2 + y)(b+1) = (ub + v)^2$$

with $x + y = u + v$.

The case $x + y = u + v = b + 1$ leads to the solution

$$(x, y, u, v; b) = (c, 3c - 2, 2c - 1, 2c - 1; 4c - 3)$$

while the case $x + y = u + v = b - 1$ leads to the equations

$$x(b+1)^2 + (b+1) = (b-1)(u+1)^2$$

or

$$x(b-1)^2 - (b-1) + (4x+2)b = (b-1)(u+1)^2.$$

Since b and $b - 1$ are coprime, it follows that for some integer m, $4x + 2 = m(b-1)$. Because $b \neq 2, 3$ and $x \leq b - 1$, $1 \leq m \leq 4$.

We obtain

$$x(b-1)^2 - (b-1) + m(b-1)b = (b-1)(u+1)^2$$

from which, after dividing by $b - 1$ and then using the substitution $b - 1 = (4x + 2)/m$, we end up with

$$(4x + 2m + 1)^2 - m(2u + 2)^2 = 1.$$

It turns out that there are no integer solutions of this diophantine equation when $m = 1, 3, 4$, so we must have $m = 2$. Thus, we have to solve $(4x + 5)^2 - 2(2u + 2)^2 = 1$. The smallest solution is $(x, y, u, v; b) = (3, 4, 5, 2; 8)$. Additional solutions are $(x, y, u, v; b) = (34, 94, 57, 71; 97)$, $(148, 142, 154, 136; 161)$.

6. See Classroom Note 252 in *Mathematical Gazette* 55 (1971), 418–419. We can take $b = n^2 + 1$, $x = (n-1)^2$ and $y = n^2$. $(xy)_b = (n^2 - n + 1)^2$.

7. We have to solve for integers the equation $x^2 - y^2 = 1111 \cdots 1111$. If the right side is written as a product ab with $a > b$, both a and b are odd and the equations $x + y = a$, $x - y = b$ can be solved for integers x and y. However, we reject any x whose square contains a zero and any pair x, y whose squares begin with the same integer.

For example, $11 = 11 \times 1$ yields $6^2 - 5^2 = 11$ and the pair $(36, 25)$; $1111 = 101 \times 11$ yields $56^2 - 45^2 = 1111$ and the pair $(3136, 2025)$; $11111 = 271 \times 41$ yields the pair $(156^2, 115^2) = (24336, 13225)$; $111111 = 481 \times 231$ yields the pair $(356^2, 125^2) = (126736, 15625)$; and $1111111 = 4649 \times$

239 yields the pair $(2444^2, 2205^2) = (5973136, 4862025)$. See *Mathematics Magazine* 59 (1986), 270–275.

8. See Problem 1234, *Mathematics Magazine* 59 (1986), 44.

9. For the bases 2 and 3, the statement is true, since $2 = (10)_2 = (2)_3$ and $2^2 = (100)_2 = (11)_3$. For bases $b \geq 4$, it is false. By an analogue of casting out nines, we see that any number is congruent to the sum of its digits (modulo $b - 1$). But $b - 1$ does not divide $2^{n+1} - 2^n = 2^n$. [Q679, *Mathematics Magazine* 55 (1982), 300, 307.]

10. The numbers in the list are of the form

$$4 \cdot 10^n(1 + 10 + \cdots + 10^{n-1}) + 8(1 + 10 + \cdots + 10^{n-1}) + 1$$

$$= \frac{4}{9} \cdot 10^n(10^n - 1) + \frac{8}{9}(10^n - 1) + 1$$

$$= \frac{4 \cdot 10^{2n} + 4 \cdot 10^n + 1}{9}$$

$$= \left(\frac{2 \cdot 10^n + 1}{3}\right)^2.$$

[Q642, *Mathematics Magazine* 49 (1976), 253, 258.]

11. If $r = 0$, the result is obvious. If $r \neq 0$, then

$$r = \left(\frac{r+6}{6}\right)^3 + \left(\frac{r-6}{6}\right)^3 + \left(\frac{-r}{6}\right) + \left(\frac{-r}{6}\right).$$

[Q649, *Mathematics Magazine* 50 (1977), 266, 271.]

12. If $m = 2k + 1$, then $n^m = n(n^k)^2$. If $m = 2k$, then

$$n^m = [n^{k-1}(n - 2)]^2 + (n - 1)(2n^{k-1})^2.$$

[Q641, *Mathematics Magazine* 49 (1976), 253, 258.]

13. Problem 21, *Math Horizons*, February, 1995. Suppose that $n = r \times 10^k$ and $n^4 = s \times 10^l$ where $1 \leq r, s < 10$. We have that r^4 is the product of s and a power of 10. Thus, we are looking for r and s for which the decimal representation for s begins with a block equal to r.

First, eliminate some possibilities. It is not possible for $3 \leq r < 4$. For $3^4 = 81$ and $4^4 = 256$, so that $8 \leq s < 10$ or $1 \leq s < 3$, and s cannot begin with a block equal to r. Similarly, $5 \leq r < 6$ leads to $6 \leq s < 10$ or $1 \leq s < 2$; $6 \leq r < 7$ leads to $1 \leq s < 3$; $7 \leq r < 8$ leads to $2 \leq s < 5$; and $8 \leq r < 9$ leads to $4 \leq s < 7$. Hence $5 \leq r < 9$ is impossible. (If $r = 2k$, note that $0 < r < r + 4 \leq 4k - 1$.)

Suppose that $r = 1 + u$ where $0 < u < 1$. Then $r^4 = 1 + 4u + 6u^2 + 4u^3 + u^4$. If $s = r^4$, then $s - r = 3u + 6u^2 + 4u^3 + u^4 > u$ and so s cannot begin with a block equal to u. If $10s = r^4$, then

$$r - s = 0.9 + 0.6u - 0.6u^2 - 0.4u^3 - 0.1u^4 > 0$$

which is not permitted. Hence $1 \leq r < 2$ is impossible.

Suppose that $r = 9 + u$ where $0 < u < 1$. Then $6561 \leq r^4 < 10000$, so that $6.561 < s = r^4 \times 10^{-3} < 10$. We have that

$$r^4 = 6561 + 2916u + 486u^2 + 36u^3 + u^4$$

whence $s = 6.561 + 2.916u + 0.486u^2 + 0.036u^3 + 0.001$.   Therefore

$$r - s = 2.439 - 1.916u - 0.486u^2 - 0.036u^3 - 0.001u^4$$

$$> 2.439(1 - u) > 0.$$

Thus, $9 \leq r < 10$ is not possible.

We conclude that $2 \leq r < 3$ or $4 \leq r < 5$. Let $r = 2 + u$ with $0 < u < 1$. Then $r^4 = 16 + 32u + 24u^2 + 8u^3 + u^4$ and $s = 1.6 + 3.2u + 2.4u^2 + 0.8u^3 + 0.1u^4$. Hence

$$s - r = f(u) \equiv 2.2u + 2.4u^2 + 0.8u^3 + 0.1u^4 - 0.4.$$

We need to select u to make $f(u)$ very small. When $0 \leq x \leq 1$, $f(x)$ is an increasing function that vanishes only near $x = 0.1544347$. Then, r should be near 2.1544347. We now use trial and error on n. If $n = 21$, then $n^4 = 199481$ while if $n = 22$, then $n^4 = 234256$. Similarly, $n = 215$, 216, 2154, 2155, 21544, 21545, 215443, 215444, 2154434 all fail to work. However

$$2154435^4 = 21544359299046164137100625.$$

Let $r = 4 + u$ with $0 < u < 1$. Then $r^4 = 256 + 256u + 96u^2 + 16u^3 + u^4$ and $s = 2.56 + 2.56u + 0.96u^2 + 0.16u^3 + 0.01u^4$. Hence

$$s - r = g(u) \equiv 1.56u + 0.96u^2 + 0.16u^3 + 0.01u^4 - 1.44.$$

The polynomial $g(x)$ has a single positive root near $x = 0.64156$. Following a procedure similar to that in the last paragraph, we test $n = 46, 47, 464, 465$, 4641, 4642, 46415, 46416. All but the last fail and we get that

$$46416^4 = 4641633499322843136.$$

The desired answer is $n = 46416$.

14. Problem 4713, *American Mathematical Monthly* 63 (1956), 729; 64 (1957), 678–679. It is readily checked that $b = 2, 3, 4, 5, 8, 12, 16$ all have the required property. We show that there are no other bases. Consider several cases and assume that the property obtains.

Let $b = 4k + 1$, with $k \geq 3$. Then

$$(2k)^2 = (k - 1)(4k + 1) + (3k + 1)$$
$$(2k - 1)^2 = (k - 2)(4k + 1) + (3k + 3).$$

Since $3k + 1 < 3k + 3 \leq 4k$, $3k + 1$ and $3k + 3$ (the last digits of $(2k)^2$ and $(2k - 1)^2$ in base b) must both be squares. But it is not possible for two squares to differ by 2. When $k = 2$, we see directly that $4^2 = (17)_9$ and 7 is not a square.

Let $b = 4k + 3$, with $k \geq 1$. Then

$$(2k)^2 = (k - 1)(4k + 3) + (k + 3)$$
$$(2k + 1)^2 = k(4k + 3) + (k + 1).$$

As in the previous case, we note that the last digits cannot both be square.

Let $b = 4k + 2$, with $k \geq 1$. Then

$$(2k)^2 = (k - 1)(4k + 2) + (2k + 2)$$
$$(2k + 1)^2 = k(4k + 2) + (2k + 1).$$

Since two positive squares cannot differ by 1, the last digits cannot both be square.

Finally, let $b = 4k$, with $k \geq 5$. Suppose that $k^2 = q \cdot 4k + r$ with $0 \leq r < 4k$. Then

$$(k + 1)^2 = q \cdot 4k + r + 2k + 1$$
$$(k + 2)^2 = (q + 1)4k + r + 4$$
$$(k + 3)^2 = (q + 1)4k + r + 2k + 9.$$

Now,

$$0 < r + 4 \leq 4k - 1 + 4 = 4k + 3$$
$$0 < r + 2k + 1 < r + 2k + 9 \leq 4k - 1 + 2k + 9 = 6k + 8.$$

If $0 \leq r + 2k + 9 \leq 4k - 1$, then $r + 2k + 1$ and $r + 2k + 9$ are squares differing by 8. But this is impossible, since the only squares differing by 8 are 1 and 9, and k is positive.

If $0 \leq 4k \leq r + 2k + 1$, then

$$(k + 1)^2 = (q + 1)4k + r - 2k + 1$$
$$(k + 3)^2 = (q + 2)4k + r - 2k + 9$$

with $0 \leq r - 2k + 9 \leq 4k - 1 - 2k + 9 = 2k + 8 \leq 4k - 1$, so that again this is impossible. (If $r = 2k$, note that $0 < r < r + 4 \leq 4k - 1$.)

Now, let $r + 2k + 1 \leq 4k - 1 \leq 4k \leq r + 2k + 9$, so that $r + 4 \leq 2k + 2 < 4k$. Since r is divisible by k, $r \geq 5$. The last digits of k^2 and $(k+2)^2$ are r and $r+4$, two squares that differ by 4. But this is not possible under the circumstances. The problem is solved.

15. Problem E1241, *American Mathematical Monthly* 63 (1956), 723; 64 (1957), 432–433. There are two solutions

$$27556 \quad 3249 \quad 81 \quad 400$$
$$34225 \quad 7396 \quad 81 \quad 900.$$

16. Problem 3935, *American Mathematical Monthly* 46 (1939), 656; 48 (1941), 344–345.

(a) Let b be the base and a the digit. Then a and $a(b^2 + 1)(b+1)$ must both be square. Hence $(b^2 + 1)(b + 1)$ is a square. Since $b \geq 2$, it is not possible for $b^2 + 1$ to be a square, so $b^2 + 1$ and $b + 1$ are not relatively prime. Therefore, $b^2 + 1$ and $b + 1$ have greatest common divisor 2, and $b^2 + 1 = 2u^2$ and $b + 1 = 2v^2$ for some positive integers u and v. This leads to $u^2 = v^4 + (v^2 - 1)^2$. One solution is $(u, v) = (5, 2)$, leading to $b = 7$ and $(1111)_7 = 20^2$ and $(4444)_7 = 40^2$. It turns out that this is the only solution.

(b) $(aaaa)_b$ is a square when $(a, b) = (1, 7), (4, 7), (21, 41)$. In general, suppose that $b^2 + 1 = 2r^2$ for some r. Then b is odd and of the form $2s - 1$ where $s < b$. Take $a = s$. Besides 7 and 41, possible values of b include 239, 1393, and 8119.

17. Problem E548, *American Mathematical Monthly* 50 (1943), 513.

(a) The squares of 1692, 2878, 2978, 4738, 5162, 6668, 9068, 9092 satisfy the requirement.

(b) Observe that, with k sixes,

$$(66\cdots68)^2 = \left[6(10^k + \cdots + 10) + 8\right]^2 = \left[6 \cdot 10 \cdot \frac{10^k - 1}{9} + 8\right]^2$$

$$= \left[\frac{2 \cdot 10^{k+1} + 4}{3}\right]^2 = \frac{1}{9}\left[4 \cdot 10^{2(k+1)} + 16 \cdot 10^{k+1} + 16\right]$$

$$= \frac{4(10^{2k+2} - 1) + 16(10^{k+1} - 1) + 36}{9}$$

$$= 4(10^{2k+1} + 10^{2k} + \cdots + 10^k + \cdots + 1)$$

$$\quad + 16(10^k + \cdots + 1) + 4$$

$$= 4(10^{2k+1} + \cdots + 10^{k+1}) + 2(10^{k+1} + \cdots + 10) + 4$$

$$= 4\ldots462\ldots24.$$

For example, $68^2 = 4624$, $668^2 = 446224$, $6668^2 = 44462224$.

(c) $2^2 = 4$, $8^2 = 64$.

(d) Since even numbers have even squares, any counterexamples must be odd. Observe that

$$(10a + c)^2 = 100a^2 + 20ac + c^2.$$

Since the carry from the square of an odd digit is 0, 2, 4, or 8, the tens digit must be even.

18. Problem E592, *American Mathematical Monthly* 50 (1943), 560; 51 (1944), 288–289.

(a) For positive integer n,

$$7 \cdot 10^{2n} + (10^{2n-1} + \cdots + 10^{n+1}) + 2 \cdot 10^n + 8(10^{n-1} + \cdots + 10) + 9$$

$$= 7 \cdot 10^{2n} + \left(\frac{10^{n-1} - 1}{9}\right) \cdot 10^{n+1} + 2 \cdot 10^n$$

$$+ 80\left(\frac{10^{n-1} - 1}{9}\right) + 9$$

$$= \frac{1}{9}\left[64 \cdot 10^{2n} + 16 \cdot 10^n + 1\right] = \left[\frac{8 \cdot 10^n + 1}{3}\right]^2.$$

(c) $(b - 1)(b^{4n-1} + \cdots + b^{3n+1}) + (b - 3)b^{3n}$

$$+ (b - 1)(b^{3n-1} + \cdots + b^{2n+1}) + (b - 3)b^{2n} + 4b^n + 4$$

$$= b^{4n} - b^{3n+1} + b^{3n+1} - 3b^{3n} + b^{3n} - b^{2n+1} + b^{2n+1}$$

$$- 3b^{2n} + 4b^n + 4$$

$$= b^{4n} - 2b^{3n} - 3b^{2n} + 4b^n + 4 = (b^{2n} - b^n - 2)^2.$$

19. Observe, for $0 \le x \le 12$, we have that

$$(956 + x)^2 = (913 + 2x)10^3 + (936 - 88x + x^2)$$

with $0 < 936 - 88x + x^2 < 1000$. Adding the numbers formed by the first three and the last three digits yields

$$(913 + 2x) + (936 - 88x + x^2) = (43 - x)^2.$$

CHAPTER **7**

Interesting Sets

Sets of numbers with interesting properties

$\{7, 18\}$ $7^3 = 1 + 18 + 18^2$ **(1)**.

$\{7, 20\}$ The sum of the divisors of 7^3 is equal to 20^2,

$\{18, 1746\}$ The sum is the square of 42 and the difference is the square of 12 **(2)**.

$\{120, 168\}$ If the sum and the difference are both increased by 1, the results are squares **(3)**.

$\{273, 364\}$ The sum of the squares and the sum of the cubes of these numbers are both squares (of 455 and 8281) **(4)**.

$\{1729, 43098\}$ The sum of the divisors of 43098^2 is equal to 1729^3 **(5)**.

$\{1061652293520, 4565486027761\}$ The sum of these numbers is the square of 2372159 while the sum of their squares is the fourth power of 2165017 **(6)**.

$\{1, 2, 3\}$ $1^2 + 2^2 + 3^2 = 2 \times 7$ and $1^4 + 2^4 + 3^4 = 2 \times 7^2$.

$\{4, 60, 105\}$ The sum of these numbers is the square of 13 and the sum of their squares is the fourth power of 11.

$\{9, 15, -16\}$ The sum and the sum of cubes are both 2^3.

$\{13, 40, 45\}$ The square of the sum of any two minus the square of the third is a square **(7)**.

$\{36, 37, -46\}$ The sum is 3^3 while the sum of cubes is $(-3)^3$.

$\{54, 72, 90\}$ The sum of these numbers and the sum of their cubes are both cubes, namely 6^3 and $108^3 = 3^3 \times 6^6$. Also $54^2 + 72^2 = 90^2$.

$\{57, 112, 672\}$ The sum of any pair and the sum of all three of the numbers are squares (of 13, 27, 28, 29). **(8)**

$\{64, 152, 409\}$ The sum of these numbers and the sum of their squares are both fourth powers (of 5 and 21).

$\{124, 957, 13852800\}$ This is the set $\{a, b, c\}$ with the smallest positive integers for which $a^2 + b^2$, $a^2 + c$, $b^2 + c$, and $a^2 + b^2 + c$ are all squares (of 965, 3724, 3843, 3845) **(9)**.

$\{482, 3362, 6242\}$ These three numbers are in arithmetic progression and the sum of any pair is a square (of 62, 82, and 98) **(10)**.

$\{567, 1008, 1792\}$ These three numbers are in geometric progression and the differences between pairs of them are squares of three numbers 21, 28, 35 in arithmetic progression.

$\{23409, 34225, 485809\}$ The numbers are the squares of 153, 185, and 697, and the differences of pairs of them are the squares of 104, 672, and 680. We can use these figures to derive other triples with the same property, such as $\{672^2, 680^2, 697^2\}$.

$\{38416, 76832, 115248\}$ The sum of the squares of these numbers is 2744^3 and the sum of their cubes is 45177216^2 **(11)**.

$\{150568, 420968, 434657\}$ The sums and the differences of all pairs of these numbers are squares (of 756, 765, 925, 520, 533, 117) **(12)**.

$\{856350, 949986, 993250\}$ All sums and differences of pairs of these numbers are squares; the sums are 1344^2, 1360^2, 1394^2 and the differences are 306^2, 370^2, 208^2.

$\{1873432, 2288168, 2399057\}$ The sum and the difference of any pair of these numbers are all squares (of 2040, 2067, 2165, 644, 725, 333) **(12)**.

$\{273267248670, 412095790665, 194479084666\}$ The sum, the sum of squares, and the sum of cubes of these numbers are all squares (of 937999, 531337975541, 312642220500391339) **(13)**.

$\{1, 22, 41, 58\}$ Any three of these numbers add up to a perfect square.

$\{5, 7, 11, 13\}$ is the only set of four consecutive primes whose average is a square **(14)**.

$\{120, 177, 232, 432\}$ The sums of any three and of all four are perfect squares.

$\{136, 264, 384, 441\}$ Seven of the fifteen nonvoid subsets of this quartet sum to perfect squares. We have that $441 = 21^2$, $136 + 264 = 20^2$, $136 + 264 + 384 = 28^2$, $136 + 264 + 441 = 29^2$, $136 + 384 + 441 = 31^2$, $264 + 384 + 441 = 33^2$ and $136 + 264 + 384 + 441 = 35^2$.

$\{585, 2016, -416, 5040\}$ The sum of any pair and the sum of all four of these numbers are squares **(15)**.

$\{1873432, 2288168, 2399057, 6560657\}$ The six differences of pairs of these numbers are the squares of 644, 725, 2165, 333, 2067, 2040.

$\{7442, 28658, 148583, 177458, 763442\}$ The sums of the pairs of these numbers are the squares of 190, 395, 430, 878, 421, 454, 890, 571, 955, and 970 **(16)**.

$\{26072323311568661931, 43744839742282591947,$
   $118132654413675138222, 186378732807587076747,$
   $519650114814905002347\}$ Stan Wagon, using *Mathematica*, discovered this set of five numbers, any three of which add up to a perfect square. The roots of these squares are 13709479110, 16006120575, 24278947215, 18181961130, 25765385550, 27057368145, 18661624446, 26106083754, 27381995679, 28708213146.

$\{-15863902, 17798783, 21126338, 49064546, 82221218, 447422978\}$ The sums of the pairs of these numbers are the squares of 1391, 2294, 5762, 8146, 20774, 6239, 8177, 10001, 21569, 8378, 10166, 21646, 11458, 22282, 23014 **(16)**.

Products differing from a square by 1

We begin our story with the Fibonacci sequence

$$1, 1, 2, 3, 5, 8, 13, 21, 34, 55, 89, 144, 233, 377, 610, \ldots$$

in which each term after the second is the sum of its two predecessors. By taking alternate terms, we form two classes of triples of numbers:

   I : $\{1, 3, 8\}$, $\{3, 8, 21\}$, $\{8, 21, 55\}$ $\{21, 55, 144\}$, $\{55, 144, 377\}, \ldots$

   II : $\{1, 2, 5\}$, $\{2, 5, 13\}$, $\{5, 13, 34\}$, $\{13, 34, 89\}$, $\{34, 89, 233\}, \ldots$.

If $\{a, b, c\}$ belongs to Class I, then $ab + 1$, $ac + 1$, and $bc + 1$ are all squares. If $\{p, q, r\}$ belongs to Class II, then $pq - 1$, $pr - 1$, and $qr - 1$ are all squares **(17)**.

For example,

$$8 \times 21 + 1 = 13^2, \quad 8 \times 55 + 1 = 21^2, \quad 21 \times 55 + 1 = 34^2,$$

and

$$5 \times 13 - 1 = 8^2, \quad 5 \times 34 - 1 = 13^2, \quad 13 \times 34 - 1 = 21^2.$$

(Do you notice anything about the root of these squares?)

Thus, we see that there are lots of triples $\{a, b, c\}$ for which $ab + 1$, $ac + 1$, and $bc + 1$ are all squares. Indeed, we can find parametric families:

$\{a, b, c\}$	$ab + 1$	$ac + 1$	$bc + 1$
$\{k - 1, k + 1, 4k\}$	k^2	$(2k - 1)^2$	$(2k + 1)^2$
$*\{1, (k - 1)(k + 1), k(k + 2)\}$	k^2	$(k + 1)^2$	$(k^2 + k - 1)^2$
$\{2, 2k(k + 1), 2(k + 1)(k + 2)\}$	$(2k + 1)^2$	$(2k + 3)^2$	$(2k^2 + 4k + 1)^2$
$\{3, k(3k + 2), (k + 1)(3k + 5)\}$	$(3k + 1)^2$	$(3k + 4)^2$	$(3k^2 + 5k + 1)^2$
$\{3, k(3k - 2), (k + 1)(3k + 1)\}$	$(3k - 1)^2$	$(3k + 2)^2$	$(3k^2 + k - 1)^2$
$*\{4, k(k + 1), (k + 2)(k + 3)\}$	$(2k + 1)^2$	$(2k + 5)^2$	$(k^2 + 3k + 1)^2$

Note that the quantity $bc + 1$ for the two triples marked with an asterisk illustrate the fact that the product of any four consecutive integers increased by 1 is a perfect square.

It is remarkable that, if we have two numbers a and b for which $ab + 1$ is square, then we can *always* find a number c for which $ac + 1$ and $bc + 1$ are both square **(18)**. The reader may wish to experiment before proceeding. For example, look at all the possible pairs $\{a, b\}$ for which $ab + 1 = 5^2$ or $ab + 1 = 11^2$, and see whether you can augment them by a suitable value of c.

The proof is surprisingly simple. Suppose that $ab + 1 = m^2$. Set $c = a + b + 2m$. Then

$$ac + 1 = a^2 + ab + 2am + 1 = a^2 + 2am + m^2 = (a + m)^2$$

$$bc + 1 = ab + b^2 + 2bm + 1 = b^2 + 2bm + m^2 = (b + m)^2.$$

Similarly, if $pq - 1$ is square, then we can find r so that $pr - 1$ and $qr - 1$ are square. Just take $r = p + q + 2n$ where $pq - 1 = n^2$. Some sets $\{p, q, r\}$ for which the product of any two less 1 is square are $\{1, k^2 + 1, (k + 1)^2 + 1\}$ and $\{2, (k - 1)^2 + k^2, k^2 + (k + 1)^2\}$.

To carry the investigation further, we may ask for sets of four integers $\{a, b, c, d\}$ for which the product of any pair plus 1 is square. One example is $\{1, 3, 8, 120\}$. This can be generalized to $\{n - 1, n + 1, 4n, 4n(4n^2 - 1)\}$ or to $\{1, n^2 - 1, n(n + 2), 4n(n^3 + 2n^2 - 1)\}$ where n is any integer. Even more general is the set

$$\{m, n^2 - 1 + (m - 1)(n - 1)^2, n(mn + 2),$$
$$4m^3 n^4 + 8m^2(2 - m)n^3 + 4m(m - 1)(m - 5)n^2$$
$$+ 4(2m - 1)(m - 2)n + 4(m - 1)\}$$

(19). S. T. Thakar extends the foregoing process for generating triples to obtain quadruples $\{a, b, c, d\}$ for which each pairwise product is one less than a square. Suppose $ab + 1 = m^2$. Then let $c = a + b + 2m$ and

$d = 4m(a + m)(b + m)$. He also observes that $\sqrt{ad + 1} + \sqrt{bd + 1} + 1 = \sqrt{cd + 1}$. If $a > m > b$, then we can also take $c = a + b - 2m$ and $d = 4m(a - m)(m - b)$ **(20)**. In 1968, the Dutch mathematician, Van Lint, asked whether there was any number n besides 120 for which the set $\{1, 3, 8, n\}$ had this property. By a very delicate argument in diophantine analysis, Baker and Davenport were able to give a negative answer; subsequently, a more elementary argument was given **(21)**.

In the same spirit, let us ask for three integers x, y, z for which $xy + z$, $yz + x$, $xz + y$ are all square **(22)**. Such triples are surprisingly abundant. Take any integer a and select u and v so that $uv = a^2 + 1$. Then

$$(x, y, z) = (u - 1, v - 1, 2a + u + v - 1)$$

works. Indeed,

$$xy + z = (a + 1)^2, \quad yz + x = (a + v - 1)^2, \quad \text{and} \quad zx + y = (a + u - 1)^2.$$

As a bonus, we also have that $xy + x + y = a^2$, $zx + z + x = (u + a)^2$ and $yz + y + z = (v + a)^2$.

Notes

1. Hugh M. Edgar discusses some open questions concerning the diophantine equation $1 + a + a^2 + \cdots + a^{x-1} = p^y$ in *American Mathematical Monthly* 81 (1974), 758–759. See also Problem 274 in *American Mathematical Monthly* 24 (1917), 467.

2. See Problem 4217 in *American Mathematical Monthly* 53 (1946), 471; 55 (1948), 36.

3. See *American Mathematical Monthly* 1 (1894), 325–326. Another such pair is $\{960, 1155\}$.

4–5. See L. E. Dickson, *History of the Theory of Numbers, Volume I*, page 54, seq. for more on this.

6. These numbers are given (incorrectly) in *American Mathematical Monthly* 11 (1904), 39. In *American Mathematical Monthly* 10 (1903), 272, the pair $\{120, -119\}$ is also given; the sum is a square and the sum of the squares is a fourth power.

7. *American Mathematical Monthly* 1 (1894), 361–362.

8. *American Mathematical Monthly* 10 (1903), 141–143. In the solution to Problem 254 (*American Mathematical Monthly* 23 (1916), 342; 24 (1917), 393), the impossibility of having three squares for which the sum of any pair and of all three is a square is demonstrated.

9. Note 2918 in *Mathematical Gazette* 44 (1960), 219–220 gives a method of finding infinitely many such sets $\{a, b, c\}$.

10. *American Mathematical Monthly* 1 (1894), 96.

11. *American Mathematical Monthly* 1 (1894), 363.

12. The problem of finding three numbers whose pairwise sums and differences are all squares was notorious in the seventeenth century and was solved by several mathematicians. An historical discussion of this problem and some methods of solutions appears in P. Nastasi and A. Scimone, Pietro Mengoli and the six-square problem, *Historia Mathematica* 21 (1994), 10–27.

Mengoli had the misfortune of having an erroneous proof of the unsolvability of this problem exposed by Jacques Ozanan, who provided the solution given here. Returning to the problem, Mengoli obtained the solution $\{u, v, w\}$, where

$$u = \frac{1}{2}\left[(p^2t^2 + s^2q^2) - (p^2q^2 + s^2t^2)\right]$$
$$v = u + (pq - st)^2 \qquad w = v + 4pstq$$

and $pstq$, $p^2 + s^2$ and $q^2 + t^2$ are to be made square. For example, one can take $(p, s, q, t) = (112, 15, 35, 12)$ or $(364, 27, 84, 13)$.

More recently, the problem was considered by R. C. Lyness in Note 2952 in *Mathematical Gazette* 45 (1961), 207–209. Suppose $b+c = p^2$, $c+a = q^2$, $a + b = r^2$, $c - b = u^2$, $c - a = v^2$ and $b - a = w^2$. Then

$$p^2 - q^2 = w^2, \quad q^2 - r^2 = u^2, \quad p^2 - r^2 = v^2. \tag{$*$}$$

On the other hand, if p, q, r, u, v, w are chosen to satisfy $(*)$, then let $a = \frac{1}{2}(q^2 - v^2)$, $b = \frac{1}{2}(p^2 - u^2)$, $c = \frac{1}{2}(p^2 + u^2)$. Noting that $v^2 = u^2 + w^2$, we find that the sums and differences of pairs of a, b, c are squares as desired.

13. *American Mathematical Monthly* 9 (1902), 145–146.

14. Q544, *Mathematics Magazine* 45 (1972), 168 asks for a proof of this fact.

15. Finding such a set is not a matter of looking for a needle in a haystack. If x, y, z, w are the four numbers, we want $x+y = a^2$, $z+w = b^2$, $x+z = c^2$, $y+w = d^2$, $x+w = e^2$ and $y+z = f^2$ for some integers a, b, c, d, e, f. Thus we need $a^2 + b^2 = c^2 + d^2 = e^2 + f^2$. Use the pythagorean triples $(3, 4, 5)$, $(8, 15, 17)$, and $(13, 84, 85)$ to obtain

$$51^2 + 68^2 = 40^2 + 75^2 = 13^2 + 84^2 = 85^2.$$

Take $a = 51$, etc, and solve for x, y, z, w. This method was used by Sallianne Dech of Sarnia, Ontario in an essay while she was in high school.

16. See *Mathematical Gazette* 62 (1978), 25–29. The set

$$\{30823058, 63849842, 150187058, 352514183, 1727301842\}$$

is also given. It is an open problem to determine six positive integers for which the sum of any pair is a square.

17. See note **(6)** in Chapter 3 for the way to verify these facts.

18. This was posed as a problem in the fourteenth round of the International Mathematical Talent Search, published in *Mathematics and Informatics Quarterly* 4 (1994), 153.

19. See *American Mathematical Monthly* 5 (1898), 301–302, where a different claim for the fourth member is given. Suppose m and n are positive integers, and u is an integer for which $\{m, (n-1)[m(n-1)+2], n(mn+2), u\}$ has the property that the product of any pair of its members plus 1 is a square. Then, for some x and y, we require that

$$mu + 1 = y^2 \quad \text{and} \quad (n-1)\big[m(n-1)+2\big] + 1 = x^2.$$

This leads to

$$mx^2 - (n-1)\big[m(n-1)+2\big]y^2 = m - (n-1)\big[m(n-1)+2\big]$$

or

$$z^2 - p(p+2)y^2 = m^2 - p(p+2) \tag{$*$}$$

where $z = mx$ and $p = m(n-1)$. Since the fundamental solution of $z^2 - p(p+2)y^2 = 1$ is $(z, y) = (p+1, 1)$, we can obtain new solutions for $(*)$ from known ones by the transformation

$$(z, y) \longrightarrow \big((p+1)z + p(p+2)y, z + (p+1)y\big).$$

Starting with $(z, y) = (m, 1)$, this yields in turn

$$(z, y) = \big((p+1)m + p(p+2), m + p + 1\big),$$

and

$$\big((p+1)^2 m + p(p+2)m + 2p(p+1)(p+2), 2(p+1)m + 2p^2 + 4p + 1\big)$$
$$= \big((2p^2 + 4p + 1)m + 2(p^3 + 3p^2 + p), 2(p+1)m + (2p^2 + 4p + 1)\big).$$

Using $u = (y^2 - 1)/m$ from this last solution gives the fourth member u in the text. Indeed, we can check that

$$[mn^2 + 2n]\big[4m^3 n^4 + 8m^2(2 - m)n^3 + 4m(m-1)(m-5)n^2$$
$$+ 4(2m - 1)(m - 2)n + 4(m - 1)\big] + 1$$
$$= \big[2m^2 n^3 + 2m(3 - m)n^2 - 4(m-1)n - 1\big]^2.$$

20. See *Mathematics and Informatics Quarterly* **6** (1996), 23–26. With $ab + 1 = m^2$, $c = a + b + 2m$, and $d = 4m(a + m)(b + m)$, we find that

$$ac + 1 = (a + m)^2$$
$$bc + 1 = (b + m)^2$$
$$ad + 1 = 4a^2 bm + 4a(a + b)m^2 + 4am^3 + 1$$
$$\quad = (4a^2 b + 4a^2 b + 4a)m + (4a^2 + 4ab)m^2 + 1$$
$$\quad = 4a^2 m^2 + (8a^2 b + 4a)m + 4a^2 b^2 + 4ab + 1$$
$$\quad = \big[2a(m + b) + 1\big]^2$$
$$bd + 1 = \big[2b(m + a) + 1\big]^2$$
$$cd + 1 = (a + b + 2m)\big[4abm + 4(a + b)m^2 + 4m^3\big] + 1$$
$$\quad = 8m^4 + 12(a + b)m^3 + \big[4(a + b)^2 + 8ab\big]m^2$$
$$\qquad + 4ab(a + b)m + 1$$
$$\quad = 4m^4 + 4(ab + 1)m^2 + 8(a + b)m^3 + 4(ab + 1)(a + b)m$$
$$\qquad + \big[4(a + b)^2 + 4ab\big]m^2 + 4ab(ab + 1) + 4ab(a + b)m + 1$$
$$\quad = 4m^4 + 8(a + b)m^3 + \big[4(a + b)^2 + 4(2ab + 1)\big]m^2$$
$$\qquad + 4(2ab + 1)(a + b)m + (2ab + 1)^2$$
$$\quad = \big[2m^2 + 2(a + b)m + (2ab + 1)\big]^2$$

$$= \left[2(m+a)(m+b)+1\right]^2.$$

Also

$$\sqrt{ad+1} + \sqrt{bd+1} + 1 = 2(a+b)m + 2ab + 2(ab+1) + 1$$
$$= 2m^2 + 2(a+b)m + 2ab + 1$$
$$= \sqrt{cd+1}.$$

Thakar suggests a generalization to "T-series with constant k," namely sequences of positive integers for which, given any three consecutive terms a, b, c, we have that $ab+k$, $ac+k$ and $bc+k$ are all squares. For example, $\{3, 8, 23, 59, 156, 407, 1067, 2792, \ldots\}$ is the beginning of such a sequence for $k=12$.

21. See the following two articles.

A. Baker and H. Davenport, The equations $3x^2 - 2 = y^2$ and $8x^2 - 7 = z^2$, *Quarterly Journal of Mathematics* (2) 20 (1969), 129–137.

P. Kanagasabapathy and T. Ponnudurai, The simultaneous diophantine equations $y^2 - 3x^2 = -2$ and $z^2 - 8x^2 = -7$, *Quarterly Journal of Mathematics* (2) 26 (1975), 275–278.

22. Problem 236, *American Mathematical Monthly* 24 (1917), 236.

Exercises

1. (a) Note that $2^4 = 4^2$. Are there any other pairs (x, y) of integers for which $x^y = y^x$? What if we relax the condition to require that x and y are rationals?
   (b) Determine all positive rational solutions of $x^y = y^x$ with $x > y > 0$.

2. Prove that

$$1 + 9 + 9^2 + 9^3 + \cdots + 9^n = 1 + 2 + 3 + 4 + \cdots + m,$$

where $m = 1 + 3 + 3^2 + 3^3 + \cdots + 3^n$.

3. Find three positive integers the sum of any pair of which is a cube.

4. Suppose that the greatest common divisor of any pair of the positive integers x, y, z is 1, and that

$$\frac{1}{x} + \frac{1}{y} = \frac{1}{z}.$$

Prove that $x + y$, $x - z$ and $y - z$ are all squares.

5. Determine all positive integers n for which $3^n - 8$ is a square.

6. (a) Determine all positive integers k for which $2^k + 7$ is a square.

(b) Determine as many positive integers n as you can for which $2^n - 7$ is a square.

(c) Let m be any number of the form $2^k - 1$ for $k \geq 3$. Show that there are at least two values of n for which $2^n - m$ is a square. Describe how to find all even values of n for which $2^n - m$ is a square.

7. Prove that there are just three pairs (x, y) of nonnegative integers which satisfy the equation $3 \cdot 2^x + 1 = y^2$.

8. Let n be a positive integer. What are the possible values of the greatest common divisor of $n^4 + 1$ and $(n + 1)^4 + 1$?

9. Show how to determine infinitely many pairs (a, b) of integers for which $a^2 + 1$ is a multiple of b and $b^2 + 1$ is a multiple of a. [Hint: Show that, if a and b satisfy the diophantine equation $a^2 + b^2 + 1 = 3ab$, then the required condition holds.]

Solutions

1. (a) See *Mathematics Magazine* 63 (1990), 30–33; 67 (1994), 305 (Q824).

(b) Let $x = (1 + v)y$ where v is a positive rational. Then $(1 + v)^y y^y = y^{(1+v)y}$, so that $1 + v = y^v$. It can be checked that

$$(x, y) = \left((1 + v)^{1+1/v}, (1 + v)^{1/v}\right)$$

satisfies the equation in the problem. However, in general, even when v is rational, these values of x and y are not rational.

We can obtain a rational solution by taking v to be the reciprocal $1/b$ of a positive integer b. More generally, let us try $v = a/b$, where a and b are coprime and $a \neq 1$. Then $(1 + v)^{1/v}$ is rational if and only if

$$\left(\frac{a + b}{b}\right)^{1/a} = \frac{p}{q}$$

for a coprime pair (p, q) of integers. Then $(a + b)q^a = bp^a$, whence $b = q^a$, $a + b = p^a$. It follows that

$$a = p^a - q^a \geq (q + 1)^a - q^a \geq aq.$$

Hence $q = b = 1$ and $1 + a = p^a \geq 2^a$, an impossibility when $a \neq 1$.

Hence the only rational solutions are

$$(x, y) = \left(\left(1 + \frac{1}{b}\right)^{1+b}, \left(1 + \frac{1}{b}\right)^b\right)$$

where b is a positive integer.

2. We have that

$$1 + 2 + 3 + \cdots + m = \frac{m(m+1)}{2}$$

$$= \frac{1}{2}\left(\frac{3^{n+1}-1}{2}\right)\left(\frac{3^{n+1}+1}{2}\right) = \frac{3^{2(n+1)}-1}{8}$$

$$= \frac{9^{n+1}-1}{8} = 1 + 9 + 9^2 + \cdots + 9^n.$$

See a "proof without words" of this in *Mathematics Magazine* 63 (1990), 225.

3. Let a, b, c be any three positive integers of which one or three are even and the sum of the cubes of any two of them exceeds the cube of the third. Then the following three integers will fill the bill:

$$\frac{a^3+b^3-c^3}{2}, \quad \frac{a^3-b^3+c^3}{2}, \quad \text{and} \quad \frac{-a^3+b^3+c^3}{2}.$$

4. This is Problem 3 in Round 2 of the International Mathematical Talent Search. Note that the greatest common divisor of xy and each of z and $x+y$ is 1. Since $z(x+y) = xy$, we have that $z = 1$ and $(x-1)(y-1) = 1$. Hence $(x, y, z) = (2, 2, 1)$ and the result holds.

5. Problem 1044, *Journal of Recreational Mathematics* 14 (1981–1982), 64; 15 (1982–1983), 70. Since 3^n is a square (modulo 8), n must be even, say $n = 2m$. This leads to

$$8 = 3^{2m} - x^2 = (3^m - x)(3^m + x)$$

for some x. Since the factors of the terms on the right side have the same parity, we are led to $m = 1$ and $x = \pm 1$. Hence $3^n - 8$ is square only for $n = 2$.

6. (a) The only possibility is $k = 1$. No higher value of k works since no square leaves a remainder of 3 upon division by 4.

(b) We have that $2^3 - 1 = 1^2$; $2^4 - 7 = 3^2$; $2^5 - 7 = 5^2$; $2^7 - 7 = 11^2$; and $2^{15} - 7 = 181^2$. By using advanced techniques involving the factorization of integers in $\mathbf{Q}(\sqrt{-7})$, it can be shown that $n = 3, 4, 5, 7, 15$ are the only values of the positive integer n for which $2^n - 7$ is square (see the article by Wells Johnson in *American Mathematical Monthly* 94 (1987), 59–62 or the book, L. J. Mordell, *Diophantine equations*, Academic, 1969, page 20). However, one can narrow down the possibilities using more elementary techniques. If $n = 2r$ is even and $x^2 = 2^{2r} - 7$, then $7 = (2^r + x)(2^r - x)$ forces $2^r = x + 1 = 7 - x$ from which even n can be found. For odd n, we can try a variety of approaches.

(i) $2^n - 7 \equiv 0 \pmod 5 \Leftrightarrow n \equiv 1 \pmod 4$, while $2^n - 7 \equiv 0 \pmod{25} \Leftrightarrow n = 5(4k+1)$. Thus $2^n - 7$ is divisible by 5, but not by 25 when

$n = 4k + 1$ with $k \not\equiv 1 \pmod 5$. This eliminates $n = 9, 13, 17, 21, 29, 33,$ $37, 41, \ldots$.

(ii) $2^{11} - 7 = 2041 = 13 \times 157$ is not a square.

(iii) Suppose that $2^n - 7 = x^2$ and $n = 15 + 2v$ for $v \geq 1$. Then

$$2^{15}(2^v - 1)(2^v + 1) = 2^{15}(2^{2v} - 1) = 2^n - 2^{15}$$

$$= (x - 181)(x + 181).$$

One of the factors $x - 181$ and $x + 181$ must be twice an odd number since their difference is divisible by 2 but not by 4. Hence, either $x - 181$ or $x + 181$ is divisible by 2^{14}.

If $x = 2^{14}y + 181$, then

$$x^2 = 2^{28}y^2 + 2^{15} \cdot 181y + 181^2$$

$$\Rightarrow 2^n = x^2 + 7 = 2^{15}(2^{13}y^2 + 181y + 1)$$

$$\Rightarrow 2^{2v} = 2^{13}y^2 + 181y + 1.$$

Hence, y is odd and $v \geq 7$. On the other hand, if $x = 2^{14}y - 181$, then a similar argument leads to $v \geq 7$. Thus $n \geq 31$.

This lower bound can be pushed higher by refining this type of argument.

(c) If $n = k$, $2^n - m = 1^2$ and if $n = 2(k-1)$, then $2^n - m = (2^{k-1} - 1)^2$. Suppose that $n = 2r$ and that $2^n - m = s^2$. Then

$$2^k - 1 = m = 2^{2r} - s^2 = (2^r - s)(2^r + s)$$

so that $2^r - s$ and $2^r + s$ must be codivisors of $2^k - 1$. One possibility is $2^r - s = 1$ and $2^r + s = 2^k - 1$, leading to $r = k - 1$ and $s = 2^{k-1} - 1$, a solution already given. If k is even, say $k = 2h$, then $2^k - 1 = (2^h - 1)(2^h + 1)$ and we can take $r = h$ and $s = 1$. Since the sum of two positive divisors of a whole number does not exceed the number increased by 1, we can deduce $(2^r - s) + (2^r + s) \leq (2^k - 1) + 1$, and thus $r \leq k$. Thus, to find even n for which $2^n - (2^k - 1)$ is square, we need only check n up to $2k$.

k	$2^n - (2^k - 1)$ square, n even
2	$2^2 - 3 = 1^2$
3	$2^4 - 7 = 3^2$
4	$2^4 - 15 = 1^2$, $2^6 - 15 = 7^2$
5	$2^8 - 31 = 15^2$
6	$2^6 - 63 = 1^2$, $2^{10} - 63 = 31^2$
7	$2^{12} - 127 = 63^2$

7. This is Problem E51 in *American Mathematical Monthly* 40 (1933), 423; 41 (1934), 110. The solutions $(x, y) = (0, 2), (3, 5), (4, 7)$ are quickly found. Let $x \geq 5$. Then y must have the form $6z \pm 1$ whence $2^{x-2} = z(3z \pm 1)$.

Therefore, for some nonnegative integers a and b, we have that $z = 2^a$ and $3z \pm 1 = 2^b$. Thus, $\pm 1 = 2^b - 3 \cdot 2^a$, with the result that $a = 0$ or $b = 0$. The first option is clearly impossible, while the second leads to $3 \cdot 2^{x-2} \pm 1 = 1$, again an impossibility.

8. This problem is suggested by Problem 24 in *Math Horizons*, February, 1995. Any common divisor of $n^4 + 1$ and $(n + 1)^4 + 1$ must divide their difference

$$(n + 1)^4 - n^4 = \left[(n + 1)^2 - n^2\right]\left[(n + 1)^2 + n^2\right]$$
$$= (2n + 1)(2n^2 + 2n + 1).$$

Since

$$16(n^4 + 1) = (2n + 1)(8n^3 - 4n^2 + 2n - 1) + 17$$

and

$$4(n^4 + 1) = (2n^2 + 2n + 1)(2n^2 - 2n + 1) + 3,$$

we see that any common divisor of $2n + 1$ and $n^4 + 1$ must divide 17, while any common divisor of $2n^2 + 2n + 1$ and $n^4 + 1$ must divide 3. But $2n^2 + 2n + 1$ is never divisible by 3, so $2n^2 + 2n + 1$ and $n^4 + 1$ are coprime. It follows that the only possible common divisor of $n^4 + 1$ and $(n + 1)^4 + 1$ are 1 and 17. The divisor 17 occurs if and only if $n \equiv 8 \pmod{17}$. For example, $8^4 + 1 = 17 \times 241$ and $9^4 + 1 = 17 \times 386$.

9. This problem was posed in the 1987 Tournament of the Towns. The following solution is due to A. Liu. Suppose that $(x, y) = (a, b)$ is a solution of the diophantine equation

$$x^2 + y^2 + 1 = 3xy.$$

Then a is a root of the quadratic equation

$$x^2 - 3bx + (b^2 + 1) = 0.$$

The second root is the integer $c = 3b - a$ and we find that $b^2 + c^2 + 1 = 3bc$. Thus, one solution (a, b) gives rise to a second (b, c). We now show that infinitely many solutions can be generated in this way. Suppose $a < b$; then, from the quadratic equation, we find that $c = (b^2 + 1)/a > b$. Since $(x, y) = (1, 1)$ satisfies the equation, we generate the sequence of solutions

$$(x, y) = (1, 1), (1, 2), (2, 5), (5, 13), (13, 34), \ldots.$$

Observe that (x, y) are comprised of alternate members of the Fibonacci sequence. Indeed, for any positive integer m, we find that

$$f_{2m-1}^2 + f_{2m+1}^2 - 3f_{2m-1}f_{2m+1}$$

$$= f_{2m-1}(f_{2m-1} - f_{2m+1}) + f_{2m+1}(f_{2m+1} - f_{2m-1}) - f_{2m-1}f_{2m+1}$$

$$= (f_{2m+1} - f_{2m-1})f_{2m} - f_{2m-1}f_{2m+1}$$

$$= f_{2m}^2 - f_{2m-1}f_{2m+1} = -1$$

by Note (4, a) of Chapter 3.

Appendix to Chapter 4

I. Introduction

This appendix assumes that the reader has university level mathematics. We will give more details about the vector addition method for solving Pell's equation, as it does not seem to be covered anywhere in the literature. The theory of continued fractions and its connection with Pell's equation is well known and will not be treated extensively here. In Section 2, we will indicate the relation between this and the method of getting successive approximations by adding numerators and denominators of previous under- and over-approximations. We will then apply it to the approximation of $\sqrt{d}$ and the solution of Pell's equation $x^2 - dy^2 = 1$. Later sections will treat higher degree analogues of Pell's equation.

2. Approximation of a general irrational

Let α be any positive irrational number. Suppose that $a_1 = \lfloor \alpha \rfloor$ is the greatest integer not exceeding α. We can write

$$\alpha = a_1 + \frac{1}{\alpha_1}$$

where $\alpha_1 > 1$. Continue on, defining a_n and α_n recursively by

$$a_{n+1} = \lfloor \alpha_n \rfloor \quad \text{and} \quad \alpha_n = a_{n+1} + \frac{1}{\alpha_{n+1}} \quad (n \geq 1).$$

The rational p_n/q_n is equal to the continued fraction $a_1 + 1/a_2 + 1/a_3 + 1/\cdots + 1/a_n$, where we adopt the convention that each slash has as its denominator everything which follows it. Define

$$\phi_1(a_1) - a_1$$

$$\phi_2(a_1, a_2) = a_1 a_2 + 1$$

$$\phi_n(a_1, a_2, \ldots, a_n) = a_n\phi_{n-1}(a_1, a_2, \ldots, a_{n-1}) + \phi_{n-2}(a_1, \ldots, a_{n-2})$$

for $n \geq 3$. Then it is known that $p_n = p_n(a_1, a_2, \ldots, a_n) = \phi_n(a_1, \ldots, a_n)$ and $q_n = q_n(a_1, a_2, \ldots, a_n) = \phi_{n-1}(a_2, \ldots, a_n)$ for $n \geq 2$.

Since

$$\frac{p_n(a_1, \ldots, a_n)}{q_n(a_1, \ldots, a_n)} = a_1 + \frac{1}{\dfrac{p_{n-1}(a_2, \ldots, a_n)}{q_{n-1}(a_2, \ldots, a_n)}} = a_1 + \frac{q_{n-1}(a_2, \ldots, a_n)}{p_{n-1}(a_2, \ldots, a_n)}$$

it can be seen that

$$\phi_n(a_1, a_2, a_3, \ldots, a_n) = a_1\phi_{n-1}(a_2, a_3, \ldots, a_n) + \phi_{n-2}(a_3, \ldots, a_n)$$

for $n \geq 3$, a fact that can also be obtained easily by induction from the definition of the sequence ϕ_n.

Form a sequence $\{x_m/y_m\}$ of successive rational approximations of α by taking $x_1 = \lfloor \alpha \rfloor$, $x_2 = \lceil \alpha \rceil$, and $y_1 = y_2 = 1$. For $m \geq 3$, $x_m = x_r + x_s$ and $y_m = y_r + y_s$, where x_r/y_r and x_s/y_s are the last fractions obtained so far which are respectively less than and greater than α. We have that $x_1 = a_1 = \phi_1(a_1)$ and $y_1 = 1$. The first few terms of the sequence x_m/y_m are:

$$\frac{a_1}{1}, \frac{a_1 + 1}{1}, \frac{2a_1 + 1}{2}, \ldots, \frac{ua_1 + 1}{u}, \frac{(u+1)a_1 + 1}{u+1}$$

where u is that positive integer for which

$$\frac{ua_1 + 1}{u} > \alpha > \frac{(u+1)a_1 + 1}{u+1},$$

or equivalently,

$$a_1 + \frac{1}{u} > a_1 + \frac{1}{\alpha_1} > a_1 + \frac{1}{u+1}$$

whence

$$u < \alpha_1 = a_2 + \frac{1}{\alpha_2} < u + 1.$$

Thus $u = a_2$, and $\phi_2(a_1, a_2)/\phi_1(a_2)$ belongs to the sequence x_m/y_m.

Suppose, as an induction hypothesis, it has been shown, for $k = 2, 3, \ldots, n$, that

$$\frac{\phi_{k-1}(a_1, a_2, \ldots, a_{k-1})}{\phi_{k-2}(a_2, a_3, \ldots, a_{k-1})}$$

is in the sequence, that α lies between this and each of the a_k succeeding terms in the sequence up to and including

$$\frac{\phi_k(a_1, a_2, \ldots, a_k)}{\phi_{k-1}(a_2, a_3, \ldots, a_k)}$$

and that α lies between

$$\frac{\phi_k(a_1,\ldots,a_k)}{\phi_{k-1}(a_2,\ldots,a_k)} = \frac{a_k\phi_{k-1}(a_1,\ldots,a_{k-1}) + \phi_{k-2}(a_1,\ldots,a_{k-2})}{a_k\phi_{k-2}(a_2,\ldots,a_{k-1}) + \phi_{k-3}(a_2,\ldots,a_{k-2})}$$

and

$$\frac{(a_k+1)\phi_{k-1}(a_1,\ldots,a_{k-1}) + \phi_{k-2}(a_1,\ldots,a_{k-2})}{(a_k+1)\phi_{k-2}(a_2,\ldots,a_{k-1}) + \phi_{k-3}(a_2,\ldots,a_{k-2})},$$

which are two consecutive terms in the sequence $\{x_m/y_m\}$. The next few terms of the sequence $\{x_m/y_m\}$ after $(\phi_n(a_1,\ldots,a_n))/(\phi_{n-1}(a_2,\ldots,a_n))$ are of the form

$$\psi_n(a_1, a_2, \ldots, a_n; z)$$
$$\stackrel{\text{def}}{=} \frac{(za_n+1)\phi_{n-1}(a_1,\ldots,a_{n-1}) + z\phi_{n-2}(a_1,\ldots,a_{n-2})}{(za_n+1)\phi_{n-2}(a_2,\ldots,a_{n-1}) + z\phi_{n-3}(a_2,\ldots,a_{n-2})}$$
$$= \frac{z\phi_n(a_1, a_2, \ldots, a_n) + \phi_{n-1}(a_1, a_2, \ldots, a_{n-1})}{z\phi_{n-1}(a_2, a_3, \ldots, a_n) + \phi_{n-2}(a_2, a_3, \ldots, a_{n-1})}$$

where $z = 1, 2, 3, \ldots, u, u+1$, and for $1 \le z \le u$, the fraction $\psi_n(z)$ lies on the same side of α while for $z = u+1$ it lies on the other side of α.

For the sake of argument, we can take n to be odd (the even case can be handled similarly). Thus

$$\psi_n(a_1, a_2, \ldots, a_n; u) > \alpha > \psi_n(a_1, a_2, \ldots, a_n; u+1).$$

Since $\phi_{n-1}(a_1, \ldots, a_{n-1}) = a_1\phi_{n-2}(a_2, \ldots, a_{n-1}) + \phi_{n-3}(a_3, \ldots, a_{n-1})$ and $\phi_{n-2}(a_1, \ldots, a_{n-2}) = a_1\phi_{n-3}(a_2, \ldots, a_{n-2}) + \phi_{n-4}(a_3, \ldots, a_{n-2})$, we find that

$$a_1 + \frac{(ua_n+1)\phi_{n-3}(a_3, \ldots, a_{n-1}) + u\phi_{n-4}(a_3, \ldots, a_{n-2})}{(ua_n+1)\phi_{n-2}(a_2, \ldots, a_{n-1}) + u\phi_{n-3}(a_2, \ldots, a_{n-2})}$$
$$> \alpha = a_1 + \frac{1}{\alpha_1}$$
$$> a_1 + \frac{(\overline{u+1}a_n+1)\phi_{n-3}(a_3, \ldots, a_{n-1}) + (u+1)\phi_{n-4}(a_3, \ldots, a_{n-2})}{(\overline{u+1}a_n+1)\phi_{n-2}(a_2, \ldots, a_{n-1}) + (u+1)\phi_{n-3}(a_2, \ldots, a_{n-2})}$$

whence

$$\psi_{n-1}(a_2, \ldots, a_n; u) < \alpha_1 < \psi_{n-1}(a_2, \ldots, a_n; u+1).$$

We continue on in this way to find that, for $1 \le l \le n-3$, $\alpha_{l-1} = a_l + 1/\alpha_l$ lies between

$$\psi_{n+1-l}(a_l, \ldots, a_n; u) \quad \text{and} \quad \psi_{n+1-l}(a_l, \ldots, a_n; u+1).$$

With $l = n - 3$, the left inequality reads

$$\frac{(ua_n + 1)\phi_3(a_{n-3}, a_{n-2}, a_{n-1}) + u\phi_2(a_{n-3}, a_{n-2})}{(ua_n + 1)\phi_2(a_{n-2}, a_{n-1}) + u\phi_1(a_{n-2})} < \alpha_{n-4}$$

$$= a_{n-3} + \frac{1}{\alpha_{n-3}}$$

or

$$\frac{(ua_n + 1)[a_{n-3}(a_{n-2}a_{n-1} + 1) + a_{n-1}] + u[a_{n-3}a_{n-2} + 1]}{(ua_n + 1)[a_{n-2}a_{n-1} + 1] + ua_{n-2}}$$

$$< a_{n-3} + \frac{1}{\alpha_{n-3}} \implies \frac{(ua_n + 1)a_{n-1} + u}{(ua_n + 1)[a_{n-2}a_{n-1} + 1] + ua_{n-2}} < \frac{1}{\alpha_{n-3}}$$

$$\implies \frac{(ua_n + 1)(a_{n-2}a_{n-1} + 1) + ua_{n-2}}{(ua_n + 1)a_{n-1} + u} > \alpha_{n-3}$$

$$= a_{n-2} + \frac{1}{\alpha_{n-2}}$$

$$\implies \frac{(ua_n + 1)a_{n-1} + u}{ua_n + 1} < \alpha_{n-2} = a_{n-1} + \frac{1}{\alpha_{n-1}}$$

$$\implies a_n + \frac{1}{u} > \alpha_{n-1} = a_n + \frac{1}{\alpha_n} \implies u < \alpha_n.$$

The right inequality for the $l = n - 3$ case leads similarly to $\alpha_n < u + 1$. Hence, we must have $u = a_{n+1}$. Thus the induction hypothesis holds for $k = n + 1$.

We conclude that the sequence $\{\alpha - \frac{x_m}{y_m}\}$ has one negative term followed by a_2 positive terms, followed by a_3 negative terms, and so on.

3. Pell's equation

We specialize to the case $\alpha = \sqrt{d}$, where d is a positive nonsquare integer. For $n \geq 3$ and $0 \leq z \leq a_{n+1} + 1$, define

$$x(z) = z\phi_n(a_1, \ldots, a_n) + \phi_{n-1}(a_1, \ldots, a_{n-1})$$

$$y(z) = z\phi_{n-1}(a_2, \ldots, a_n) + \phi_{n-2}(a_2, \ldots, a_{n-1}).$$

Then

$$x(z)^2 - dy(z)^2 = \left[\phi_n^2(a_1, \ldots, a_n) - d\phi_{n-1}^2(a_2, \ldots, a_n)\right]z^2$$

$$+ 2\left[\phi_n(a_1, \ldots, a_n)\phi_{n-1}(a_1, \ldots, a_{n-1})\right.$$

$$\left. - d\phi_{n-1}(a_2, \ldots, a_n)\phi_{n-2}(a_2, \ldots, a_{n-1})\right]z$$

$$+ \left[\phi_{n-1}^2(a_1, \ldots, a_{n-1}) - d\phi_{n-2}^2(a_2, \ldots, a_{n-1})\right].$$

For $0 \leq z \leq a_{n+1}$, these are consecutive values *of the same sign* in the sequence $\{x_m^2 - dy_m^2\}$, and for $1 \leq z \leq a_n$ actually consecutive terms.

The next term in the sequence is given by $z = a_n + 1$ and has opposite sign. These $a_n + 1$ values are consecutive values of a quadratic polynomial, whose second-order difference is $2[\phi_n^2(a_1, \ldots, a_n) - d\phi_{n-1}^2(a_2, \ldots, a_n)]$.

From the foregoing results, we can determine an algorithm to generate all the values of $w_m = x_m^2 - dy_m^2$ solely from the knowledge of $x_1^2 - dy_1^2$ and $x_2^2 - dy_2^2$, without it being necessary to determine x_m and y_m individually. Then it is possible to recreate the continued fraction table for the solutions of $x^2 - dy^2 = 1$.

Suppose that k is a positive integer for which $k < \sqrt{d} < k + 1$ and that $-a = x_1^2 - dy_1^2 = k^2 - d$, $b = x_2^2 - dy_2^2 = (k+1)^2 - d$. Then $(x_3, y_3) = (2k + 1, 2)$, so that

$$x_3^2 - dy_3^2 = 4k^2 + 4k + 1 - 4d = 2(b - a) - 1$$

is easily found from the first two values $-a$ and b of $x_m^2 - dy_m^2$.

Suppose that we have found the values of w_m for $1 \leq m \leq s$. Let w_r and w_s be the last two terms of the sequence with one sign and w_t the last term with the opposite sign. Form the difference table

w_r		w_s		$2w_s - w_r + 2w_t$
	$w_s - w_r$		$w_s - w_r + 2w_t$	
		$2w_t$		

where each row is the set of differences of the adjacent elements of its predecessors. We will call this the *quadratic algorithm* since each block of consecutive values of a given sign will be consecutive values of a quadratic polynomial whose second-order difference is fixed at twice a certain entry in the sequence.

It is best to illustrate by an example. Take $d = 54$ so that $-a = 7^2 - 54 = -5$ and $b = 8^2 - 54 = 10$. The sequence of w_m begins with the terms $-5, 10, 2(10 - 5) - 1 = 9$.

Form the table

10		9		-2
	-1		-11	
		-10		

where we enter the terms in the following order:

i. 10 and 9, being the last two positive terms so far;

ii. -1, being the difference $9 - 10$;

iii. -10, being twice the last negative term so far;

iv. -11, being that number from which -1 is subtracted to give the difference -10;

v. -2, being that number from which 9 is subtracted to given the difference -10.

The terms w_m obtained so far are -5, 10, 9, -2. For the next leg we have the table

$$
\begin{array}{ccccc}
-5 & & -2 & & 19 \\
 & 3 & & 21 & \\
 & & 18 & &
\end{array}
$$

extending the sequence to -5, 10, 9, -2, 19.

The next table yields a new term of the same sign. Since the second-order difference will not change until a sign change occurs, we will continue the table until this happens:

$$
\begin{array}{ccccccccccccccc}
9 & & 19 & & 25 & & 27 & & 25 & & 19 & & 9 & & -5 \\
 & 10 & & 6 & & 2 & & -2 & & -6 & & -10 & & -14 & \\
 & & -4 & & -4 & & -4 & & -4 & & -4 & & -4 & &
\end{array}
$$

The sequence is extended to

$$-5,\ 10,\ 9,\ -2,\ 19,\ 25,\ 27,\ 25,\ 19,\ 9,\ -5.$$

We note in passing that, since the second-order difference is always of the opposite sign of the initial term of the top row of the table, the numbers in the top row will eventually change sign.

The reader can check that the next three tables are

$$
\begin{array}{ccccc}
-2 & & -5 & & 10 \\
 & -3 & & 15 & \\
 & & 18 & &
\end{array}
$$

$$
\begin{array}{ccccccc}
9 & & 10 & & 1 & & -18 \\
 & 1 & & -9 & & -19 & \\
 & & -10 & & -10 & &
\end{array}
$$

$$
\begin{array}{ccccccccccccccccc}
-5 & & -18 & & -29 & & -38 & & -45 & & -50 & & -53 & & -54 & & -53 \\
 & -13 & & -11 & & -9 & & -7 & & -5 & & -3 & & -1 & & 1 & \\
 & & 2 & & 2 & & 2 & & 2 & & 2 & & 2 & & 2 & & 2
\end{array}
$$

The sequence is now -5, 10, 9, -2, 19, 25, 27, 25, 19, 9, -5, 10, 1, -18, -29, -38, -45, -50, -53, -54, -53, ... (As a check on our work, we observe that the leg following the number 1 culminates in the number $-d = -54$.) We find that $a_1 = 7$, $a_2 = 2$, $a_3 = 1$, $a_4 = 6$, $a_5 = 1$, $a_6 = 2$, $a_7 = 14$, and we can form the table:

n	a_n	p_n	q_n	$p_n^2 - 54q_n^2$
1	7	7	1	-5
2	2	15	2	9
3	1	22	3	-2
4	6	147	20	9
5	1	169	23	-5
6	2	485	66	1
7	14	6959	947	-5

It would be interesting to have a direct proof that the sequence w_m defined recursively by $w_1 = -a$ (a negative integer), $w_2 = b$ (a positive integer), $w_3 = 2(b-a) - 1$ and w_m by the quadratic algorithm for $m > 3$, is periodic and contains the number 1, without having to go through the theory of Pell's equation.

4. The special cubic version of Pell's equation

Suppose that d is a noncube integer and let θ be its real cube root. The cubic analogue of Pell's equation is

$$g_d(x, y, z) = 1$$

where

$$g_d(x, y, z) = \frac{1}{2}(x + y\theta + z\theta^2)\big((x \quad y\theta)^2 + (y\theta - z\theta^2)^2 + (x - z\theta^2)^2\big)$$

$$= x^3 + dy^3 + d^2z^3 - 3dxyz.$$

There is no loss of generality in assuming that d is positive, as $g_d(x, y, z) = 1$ if and only if $g_{(-d)}(x, -y, z) = 1$. As in the case of the quadratic Pell's equation, there is a fundamental solution (u, v, w) from which the coefficients of 1, θ and θ^2 in the integer powers of $(u + v\theta + w\theta^2)$ provide a complete set of solutions.

By analogy with the quadratic case, we set up an algorithm to generate the nonnegative solutions. When (x, y, z) is a solution with x, y, and z large and positive, then x must be close to $y\theta$ and y must be close to $z\theta$. Accordingly, we can assign to a vector (x, y, z) the ordered pair

$$\big(\text{sign}(x^3 - dy^3), \text{sign}(y^3 - dz^3)\big).$$

This can assume the four values $(+, +), (-, -), (+, -), (-, +)$.

We begin with four vectors (x, y, z) and their associated signs:

$$\left(\lfloor d^{1/3}\lfloor d^{1/3}\rfloor\rfloor, \lfloor d^{1/3}\rfloor, 1\right) \qquad (-, -)$$

$$\left(\lceil d^{1/3}\lfloor d^{1/3}\rfloor\rceil, \lfloor d^{1/3}\rfloor, 1\right) \qquad (+, -)$$

$$\left(\lfloor d^{1/3}\lceil d^{1/3}\rceil\rfloor, \lceil d^{1/3}\rceil, 1\right) \qquad (-, +)$$

$$\left(\lceil d^{1/3}\lceil d^{1/3}\rceil\rceil, \lceil d^{1/3}\rceil, 1\right) \qquad (+, +)$$

Suppose that we have continued the table to the vector (a, b, c). Look at the sign of (a, b, c), and search through the table for the last vector (p, q, r) with the opposite signs. (Here, $(+, +)$ and $(-, -)$ are opposites, as are $(+, -)$ and $(-, +)$.) Then the next entry in the table is $(a + p, b + q, c + r)$.

Although we do not get the same regular pattern here as in the quadratic case, we find that in practice, the algorithm will surprisingly often produce the smallest positive solution to $g_d(x, y, z) = 1$. However, it does not work every time ($d = 15$ is the smallest counterexample), and when the process is continued it does not produce all the positive solutions as in the quadratic case. The tables given in the next section will provide some examples.

Besides this, there are a number of *ad hoc* methods of finding a solution to the equation $g_d(x, y, z) = 1$; a useful strategy is to go after a solution in which not all the variables are positive integers.

1. Suppose that $d = k^3 + r$ and that $rs = 3k$. Then $(x, y, z) = (1, ks, -s)$ satisfies $g_d(x, y, z)$. For this to give integer solutions, we need r to divide $3k$. However, it will always give rational solutions. For example, when $d = 29$, $r = 2$, $k = 3$ and $s = 9/2$, we get $(x, y, z) = (1, 27/2, -9/2)$; when $d = 25$, $r = -2$, $k = 3$ and $s = -9/2$, we get $(x, y, z) = (1, -27/2, 9/2)$.

2. Denote by $(x, y, z) * (u, v, w)$ the vector of coefficients of powers of θ obtained in the product $(x + y\theta + z\theta^2)(u + v\theta + w\theta^2)$. Thus,

$$(x, y, z) * (u, v, w) = (xu + dyw + dzv, xv + yu + dzw, xw + yv + zu).$$

It turns out that

$$g_d\big((x, y, z) * (u, v, w)\big) = g_d(x, y, z)g_d(u, v, w).$$

The equation $g_d(x, y, z) = 8$ is satisfied by $(x, y, z) = (2, 3a, -3)$ when $d = a^3 + 2a$. Hence,

$$(x, y, z) = (2, 3a, -3) * (2, 3a, -3) = (4 - 18ad, 12a + 9d, -12 + 9a^2)$$

satisfies $g_d(x, y, z) = 64$. If we select a so that each term in this vector is divisible by 4, we can divide each term by 4 and obtain a solution to

$g_d(x, y, z) = 1$. For example,

$a = 2$, $d = 12$, $(x, y, z) = (-107, 33, 6)$
$$\text{satisfies } x^3 + 12y^3 + 144z^3 - 36xyz = 1;$$

$a = 4$, $d = 72$, $(x, y, z) = (-1295, 174, 33)$
$$\text{satisfies } x^3 + 72y^3 + 72^2 z^3 - 216xyz = 1;$$

$a = 6$, $d = 228$, $(x, y, z) = (-6155, 531, 78)$
$$\text{satisfies } x^3 + 228y^3 + 228^2 z^3 - 684xyz = 1.$$

More generally, set $a = 2b$ so that $d = 4b(2b^2 + 1)$. Then

$$(x, y, z) = \left(1 - 36b^2(2b^2 + 1), 3b(6b^2 + 5), 3(3b^2 - 1)\right)$$

satisfies $g_d(x, y, z) = 1$.

3. Consider values of d for which there is a solution to $g_d(x, y, z) = 1$ of the form $(x, y, z) = (1, v, w)$ with $v < 0$. The condition for this is that $d = -v(v^2 - 3w)/w^3$. When $w = 1$, we have that $d = -v(v^2 - 3) = 2, 18, 52, 110, \ldots$. When $w = 2$ and v is divisible by 4, we have that $d = -v(v^2 - 6)/8 = 5, 58, 207, 500, 985, \ldots$. When $w = 3$ and v is divisible by 9, we have that $d = -v(v^2 - 9)/27$. This covers $d = 24$ and $d = 210$, for example.

4. Consider values of d for which there is a solution to $g_d(x, y, z) = 1$ of the form $(x, y, z) = (u, v, 0)$. Then we have $u^3 + dv^3 = 1$, whence $-dv^3 = u^3 - 1 = (u - 1)(u^2 + u + 1)$. If $u - 1 = -t^3$, $v = t$, then $d = u^2 + u + 1 = t^6 - 3t^3 + 3$. We have the following possibilities:

t	d	solution
1	1	$(0, 1, 0)$
-1	7	$(2 - 1, 0)$
2	43	$(-7, 2, 0)$
-2	91	$(9, -2, 0)$
3	651	$(-26, 3, 0)$
-3	813	$(28, -3, 0)$

Other solutions can be found where $u^3 - 1$ has a cubic factor. For example, when $u = 9$, $u^3 - 1 = 728 = 8 \times 91$, $-dv^3 = 2^3 \times 91$, so we can take $d = 91$, $v = -2$. When $u = 10$, $u^3 - 1 = 999 = 27 \times 37$, so we can take $d = 37$, $v = -3$.

5. Suppose that $d = r^2$. Then, since

$$u^3 + dv^3 + d^2w^3 - 3duvw = u^3 + r(rw)^3 + r^2v^3 - 3ru(rw)v,$$

$(x, y, z) = (u, v, w)$ is a solution to $g_d(x, y, z) = k$ if and only if $(x, y, z) = (u, rw, v)$ is a solution to $g_r(x, y, z) = k$.

6. Suppose that $d = r^3 s$. Then, since $u^3 + dv^3 + d^2w^3 - 3duvw = u^3 + s(rv)^3 + s^2(r^2w)^3 - 3su(rv)(r^2w)$, $(x, y, z) = (u, v, w)$ is a solution to $g_d(x, y, z) = k$ if and only if $(x, y, z) = (u, rv, r^2w)$ is a solution to $g_s(x, y, z) = k$.

7. Suppose that (x, y, z) satisfies the equation $g_d(x, y, z) = 1$. Then consideration of the expression $(x + y\theta + z\theta^2)^{-1}$ leads to the "inverse" solution $(x^2 - dyz, dz^2 - xy, y^2 - xz)$. This happens to be orthogonal as a vector to (z, y, x).

5. Tables for the cubic equations

$$x^3 + dy^3 + d^2z^3 - 3dxyz = 1$$

In the following table, we indicate the smallest positive solution (x, y, z) (the "fundamental" solution) and its inverse $(x^2 - dyz, dz^2 - xy, y^2 - xz)$, with an indication as to whether the fundamental solution arises from the algorithm of Section 4 (Y = Yes; N = No). Notice how modest in size the numbers in the inverse solution are in general.

d	fundamental solution	algorithm	inverse solution
2	$(1, 1, 1)$	Y	$(-1, 1, 0)$
3	$(4, 3, 2)$	Y	$(-2, 0, 1)$
4	$(5, 3, 2)$	Y	$(1, 1, -1)$
5	$(41, 24, 14)$	Y	$(1, -4, 2)$
6	$(109, 60, 33)$	Y	$(1, -6, 3)$
7	$(4, 2, 1)$	Y	$(2, -1, 0)$
9	$(4, 2, 1)$	Y	$(-2, 1, 0)$
10	$(181, 84, 39)$	Y	$(1, 6, -3)$
11	$(89, 40, 18)$	Y	$(1, 4, -2)$
12	$(9073, 3963, 1731)$	Y	$(-107, 33, 6)$
13	$(94, 40, 17)$	Y	$(-4, -3, 2)$
14	$(29, 12, 5)$	Y	$(1, 2, -1)$
15	$(5401, 2190, 888)$	N	$(1, -30, 12)$
16	$(16001, 6350, 2520)$	N	$(1, 50, -20)$

17	$(314892, 122465, 47628)$	Y	$(324, -252, 49)$
18	$(55, 21, 8)$	Y	$(1, -3, 1)$
19	$(12304, 4611, 1728)$	Y	$(64, -48, 9)$
20	$(391001, 144046, 53067)$	Y	$(361, -266, 49)$
21	$(1705, 618, 224)$	Y	$(-47, 6, 4)$
22	$(793, 283, 101)$	Y	$(23, 3, -4)$
23	$(2166673601, 761875860,$		
	$267901370)$	N	$(-41399, -3160, 6230)$
24	$(649, 225, 78)$	Y	$(1, -9, 3)$
25	$(9375075001, 3206230550,$		
	$1096515424)$	Y	$(70001, 13850, -12924)$
26	$(9, 3, 1)$	Y	$(3, -1, 0)$
28	$(9, 3, 1)$	Y	$(-3, 1, 0)$

The solutions to $x^3 + 2y^3 + 4z^3 - 6xyz = 1$ ($d = 2$), with a star indicating those given by the algorithm:

$(1, 0, 0)$	$(1, 0, 0)$
$(1, 1, 1)*$	$(-1, 1, 0)$
$(5, 4, 3)*$	$(1, -2, 1)$
$(19, 15, 12)*$	$(1, 3, -3)$
$(73, 58, 46)*$	$(-7, -2, 6)$
$(281, 223, 177)$	$(19, -5, -8)$
$(1081, 858, 681)$	$(-35, 24, 3)$
$(4159, 3301, 2620)*$	$(41, -59, 21)$
$(16001, 12700, 10080)$	$(1, 100, -80)$
$(61561, 48861, 38781)*$	$(-161, -99, 180)$
$(236845, 187984, 149203)$	$(521, -62, \quad 279)$
$(911219, 723235, 574032)*$	$(-1079, 583, 217)$
$(3505753, 2782518, 2208486)$	$(1513, -1662, 366)$

The algorithm for $x^3 + 15y^3 + 225z^3 - 45xyz = 1$ fails to get the smallest solution.

(x, y, z)	sign	$g_{15}(x, y, z)$
$(4, 2, 1)$	$(-, -)$	49
$(5, 2, 1)$	$(+, -)$	20
$(7, 3, 1)$	$(-, +)$	28
$(8, 3, 1)$	$(+, +)$	62
$(12, 5, 2)$	$(-, +)$	3
$(17, 7, 3)$	$(-, -)$	68
$(25, 10, 4)$	$(+, +)$	25
$(42, 17, 7)$	$(+, -)$	48

$(54, 22, 9)$	$(-, -)$	69
$(79, 32, 13)$	$(+, -)$	4
$(91, 37, 15)$	$(-, +)$	16
$(170, 69, 28)$	$(-, -)$	35
$(195, 79, 32)$	$(+, +)$	60
$(365, 148, 60)$	$(+, +)$	5
$(535, 217, 88)$	$(-, -)$	70
$(900, 365, 148)$	$(-, +)$	75
$(979, 397, 161)$	$(-, -)$	124
$(1344, 545, 221)$	$(-, -)$	84
$(1709, 693, 281)$	$(-, -)$	44
$(2074, 841, 341)$	$(-, +)$	34
$(2153, 873, 354)$	$(-, -)$	62
$(2518, 1021, 414)$	$(-, -)$	7
$(2883, 1169, 474)$	$(-, +)$	12
$(2962, 1201, 487)$	$(+, -)$	88
$(5845, 2370, 961)$		100

6. The general cubic version of Pell's equation

Pell's equation can be generalized further. Let $t^3 + at^2 + bt + c$ be a monic cubic polynomial with integer coefficients whose zeros are $\theta = \theta_1$, θ_2, and θ_3. The corresponding Pell's function is

$$g(x, y, z) = (x + y\theta_1 + z\theta_1^2)(x + y\theta_2 + z\theta_2^2)(x + y\theta_3 + z\theta_3^2)$$
$$= x^3 - cy^3 + c^2z^3 - ax^2y + (a^2 - 2b)x^2z + bxy^2 + acy^2z$$
$$+ (b^2 - 2ac)xz^2 - bcyz^2 + (3c - ab)xyz,$$

and we look for solutions of $g(x, y, z) = 1$.

Example 1. $t^3 - 7t^2 + 14t - 7$

The polynomial has three real zeros and its Pell's function is

$$g(x, y, z) = x^3 + 7y^3 + 49z^3 + 7x^2y + 21x^2z$$
$$+ 14xy^2 + 49y^2z + 98xz^2 + 98yz^2 + 77xyz.$$

The equation $g(x, y, z) = 1$ has lots of solutions. In each column, the next solution after (x, y, z) is obtained by taking the coefficients of $(1 - \theta)(x + y\theta + z\theta)$, when this product is expressed as a combination of θ and θ^2 (occurrences of higher powers of θ are replaced using the fact that θ is a root of a cubic equation).

$(29, -19, 3)$ $(-12, 7, -1)$ $(95, -63, 10)$ $(-20, 13, -2)$ $(-41, 26, -4)$ $(-66, 44, 7)$

$(8, -6, 1)$ $(-5, 5, -1)$ $(25, -18, 3)$ $(-6, 5, -1)$ $(-13, 11, -2)$ $(-17, 12, -2)$

$(1, 0, 0)$ $(2, -4, 1)$ $(4, -1, 0)$ $(1, -3, 1)$ $(1, -4, 1)$ $(-3, 1, 0)$

$(1, -1, 0)$ $(-5, 8, -2)$ $(4, -5, 1)$ $(-6, 10, -3)$ $(-6, 9, -2)$ $(-3, 4, -1)$

$(1, -2, 1)$ $(9, -15, 4)$ $(-3, 5, -1)$ $(15, -26, 8)$ $(8, -13, 3)$ $(4, -7, 2)$

$(-6, 11, -4)$ $(-19, 32, -9)$ $(4, -6, 1)$ $(-13, 21, -5)$ $(-10, 17, -5)$

$(22, -39, 13)$ $(-3, 4, 0)$ $(22, -36, 9)$

$(-69, 121, -39)$ $(-3, 7, -4)$

$(204, -356, 113)$ $(25, -46, 17)$

$(-587, 1022, -322)$

Example 2. $t^3 - t - 1$

There is one real zero and two nonreal zeros. We have

$$g(x, y, z) = x^3 + y^3 + z^3 + 2x^2 z + xz^2 - xy^2 - yz^2 - 3xyz.$$

A partial list of solutions to $g(x, y, z) = 1$ is

direct solutions	inverse solutions
$(1, 0, 0)$	$(1, 0, 0)$
$(0, 1, 0)$	$(-1, 0, 1)$
$(0, 0, 1)$	$(1, 1, -1)$
$(1, 1, 0)$	$(0, -1, 1)$
$(0, 1, 1)$	$(-1, 1, 0)$
$(1, 1, 1)$	$(2, 0, -1)$
$(1, 2, 1)$	$(-2, -1, 2)$
$(1, 2, 2)$	$(1, 2, -2)$
$(2, 3, 2)$	$(1, -2, 1)$
$(2, 4, 3)$	$(-3, 1, 1)$
$(3, 5, 4)$	$(4, 1, -3)$
$(4, 7, 5)$	$(-3, -3, 4)$
$(5, 9, 7)$	$(0, 4, -3)$
$(7, 12, 9)$	$(4, -3, 0)$
$(9, 16, 12)$	$(-7, 0, 4)$
$(12, 21, 16)$	$(7, 4, -7)$

The vector sum of two adjacent solutions gives the next solution but one, reading up the right column and down the left.

7. The special quartic version of Pell's equation

Let d be any positive nonsquare integer and θ be the real positive fourth root of d. Pell's function, in this case, is

$$g_d(x, y, z, w)$$

$$= (x + y\theta + z\theta^2 + w\theta^3)(x - y\theta + z\theta^2 - w\theta^3) \times$$

$$(x + iy\theta - z\theta^2 - iw\theta^3)(x - iy\theta - z\theta^2 + iw\theta^3)$$

$$= (x + y\theta + z\theta^2 + w\theta^3)\big[(x^3 + dy^2z + d^2zw^2 - dxz^2 - 2dxyw)$$

$$+ (dy^2w + 2dxzw - x^2y - dyz^2 - d^2w^3)\theta$$

$$+ (xy^2 + dxw^2 + dz^3 - x^2z - 2dyzw)\theta^2$$

$$+ (-y^3 - x^2w - dz^2w + dyw^2 + 2xyz)\theta^3\big]$$

$$= x^4 - dy^4 + d^2z^4 - d^3w^4 - 2dx^2z^2 + 2d^2y^2w^2$$

$$- 4dx^2yw + 4dxy^2z - 4d^2yz^2w + 4d^2xzw^2$$

$$= (x^2 + dz^2 - 2dyw)^2 - d(y^2 + dw^2 - 2xz)^2.$$

Suppose $d = 2$. Then

$$g_2(x, y, z, w) = (x^2 + 2z^2 - 4yw)^2 - 2(2xz - y^2 - 2w^2)^2.$$

Following a strategy similar to that used for the cubic case, we can form the following table for $d = 2$. The sign of (x, y, z, w) is determined by the signs of $x^4 - 2y^4$, $y^4 - 2z^4$ and $z^4 - 2w^4$.

(x, y, z, w)	sign	$x^2 + 2z^2 - 4yw$	$2xz - y^2 - 2w^2$	$g_2(x, y, z, w)$
$(1, 1, 1, 1)$	$(-,-,-)$	-1	-1	-1
$(2, 1, 1, 1)$	$(+,-,-)$	2	1	2
$(2, 2, 1, 1)$	$(-,+,-)$	-2	-2	-4
$(2, 2, 2, 1)$	$(-,-,+)$	4	2	8
$(3, 3, 2, 1)$	$(-,+,+)$	5	1	23
$(3, 2, 2, 1)$	$(+,-,+)$	-3	6	-63
$(3, 2, 1, 1)$	$(+,+,-)$	3	0	9
$(4, 3, 2, 1)$	$(+,+,+)$	12	5	94
$(5, 4, 3, 2)$	$(+,+,+)$	11	-9	-41
$(6, 5, 4, 3)$	$(+,+,+)$	8	5	14
$(7, 6, 5, 4)^*$	$(-,+,+)$	3	2	1
$(9, 7, 6, 5)$	$(+,-,+)$	13	9	7
$(11, 9, 7, 6)$	$(+,+,-)$	3	1	7
$(13, 11, 9, 7)$	$(-,+,+)$	23	15	79
$(15, 12, 10, 8)$	$(+,+,+)$	41	28	113
$(16, 13, 11, 9)$	$(+,-,+)$	30	21	18
$(18, 15, 12, 10)$	$(+,+,+)$	12	7	46
$(19, 16, 13, 11)$	$(-,+,-)$	-5	-4	-7
$(35, 29, 24, 20)$	$(+,+,+)$	57	39	207

$(36, 30, 25, 21)$	$(+, +, +)$	26	18	28
$(37, 31, 26, 22)^*$	$(+, +, -)$	-7	-5	-1
$(39, 33, 28, 23)$	$(-, -, +)$			
$(76, 64, 54, 45)$	$(-, -, +)$			
$(113, 95, 80, 67)$	$(+, -, +)$			

For the record, here are a few solutions of $g_2(x, y, z, w) = \pm 1$:

(x, y, z, w)	$g_2(x, y, z, w)$
$(1, 1, 1, 1)$	-1
$(3, 2, 2, 2)$	$+1$
$(7, 6, 5, 4)$	$+1$
$(33, 28, 24, 20)$	$+1$
$(37, 31, 26, 22)$	-1
$(195, 164, 138, 116)$	$+1$
$(1031, 867, 729, 613)$	-1
$(5449, 4582, 3853, 3240)$	$+1$

Suppose $d = 3$. Then

$$g_3(x, y, z, w) = (x^2 + 3z^2 - 6yw)^2 - 3(2xz - y^2 - 3w^2)^2.$$

The same strategy as before now gives the table:

(x, y, z, w)	sign	$x^2 + 3z^2 - 6yw$	$2xz - y^2 - 3w^2$	$g_3(x, y, z, w)$
$(1, 1, 1, 1)$	$(-, -, -)$	-2	-2	-8
$(2, 1, 1, 1)^*$	$(+, -, -)$	1	0	1
$(2, 2, 1, 1)$	$(-, +, -)$	-5	-3	-2
$(2, 2, 2, 1)$	$(-, -, +)$	4	1	13
$(3, 3, 2, 1)$	$(-, +, +)$	3	0	9
$(3, 2, 2, 1)$	$(+, -, +)$	9	5	6
$(3, 2, 1, 1)$	$(+, +, -)$	0	-1	-3
$(4, 3, 2, 1)$	$(+, +, +)$	10	4	52
$(5, 4, 3, 2)$	$(-, +, +)$	4	2	4
$(7, 5, 4, 3)^*$	$(+, -, +)$	7	4	1
$(9, 7, 5, 4)$	$(-, +, -)$	-12	-7	-3
$(16, 12, 9, 7)$	$(+, +, -)$	-5	-3	-2
$(18, 14, 11, 8)$	$(-, -, +)$	15	8	33
$(34, 26, 20, 15)$	$(-, -, +)$	16	9	13
$(50, 38, 29, 22)^*$	$(-, -, +)$	7	4	1
$(66, 50, 38, 29)$	$(+, -, -)$	-12	-7	-3
$(71, 54, 41, 31)$	$(-, +, +)$	40	23	13

Some solutions of $g_3(x, y, z, w) = 1$ can be listed

$$(2, 1, 1, 1) \qquad\qquad (7, 5, 4, 3)$$
$$(13, 10, 8, 6) \qquad\qquad (50, 38, 29, 22)$$
$$(98, 75, 57, 43)$$
$$(721, 548, 416, 316)$$
$$(5282, 4013, 3049, 2317)$$

Suppose that we have values of d, x, y for which $g_d(x, y, 0, 0) = 1$; this is equivalent to $x^4 - dy^4 = 1$. Thus, this latter equation consitutes a part of the theory of the special quartic Pell's equation.

8. The special quintic version of Pell's equation

Let d be an integer and let θ be its real fifth root. Pell's function is

$$g_d(x, y, z, u, v)$$
$$= \prod \left\{ (x + y\zeta\theta + z\zeta^2\theta^2 + u\zeta^3\theta^3 + v\zeta^4\theta^4 : \zeta^5 = 1 \right\}$$
$$= (x^5 + dy^5 + d^2 z^5 + d^3 u^5 + d^4 v^5)$$
$$\quad - 5d(x^3 yv + x^3 zu + xy^3 z)$$
$$\quad - 5d^2(y^3 uv + xz^3 v + yz^3 u + xyu^3)$$
$$\quad - 5d^3(zu^3 v + xuv^3 + yzv^3) + 5d(x^2 y^2 u + x^2 yz^2)$$
$$\quad + 5d^2(x^2 u^2 v + x^2 zv^2 + xy^2 v^2 + xz^2 u^2 + y^2 z^2 v + y^2 zu^2)$$
$$\quad + 5d^3(yu^2 v^2 + z^2 uv^2) - 5d^2(xyzuv).$$

Here are some solutions of $g_d(x, y, z, u, v) = 1$ for various values of d:

$$d = 2: (1, 1, 0, 1, 0), (1, -2, 1, 0, 0), (1, 1, 1, 1, 1), (1, 4, 1, 2, 2),$$
$$(5, 4, 4, 3, 3), (9, 8, 7, 6, 5)$$
$$d = 3: (1, 1, 0, 1, 0), (1, 5, 1, 2, 2), (7, 6, 5, 4, 3)$$
$$d = 4: (1, 1, 0, -1, 0), (9, 7, 5, 4, 3)$$
$$d = 5: (1, 0, 0, 1, -1), (76, 55, 40, 29, 21)$$

9. The special sextic version of Pell's equation

Let d be a positive integer that is not a sixth power, and let θ be its positive sixth root. Then

$$
\begin{aligned}
g_d&(x, y, z, u, v, w) \\
&= \left[x^3 + (3xu^2 + 3y^2v + z^3 - 3xyw - 3xvz - 3uyz)d \right. \\
&\qquad \left. + (v^3 + 3zw^2 - 3uvw)d^2\right]^2 \\
&\quad - d\left[(3x^2u + y^3 - 3xyz) \right. \\
&\qquad \left. + (u^3 + 3yv^2 + 3z^2w - 3xvw - 3uyw - 3uvz)d + w^3d^2\right]^2 \\
&= [x^2 + 2dzv - du^2 - 2dyw]^3 + d[2xz + dv^2 - y^2 - 2duw]^3 \\
&\quad + d^2[z^2 + 2xv - 2yu - dw^2]^3 - 3d[x^2 + 2dzv - du^2 - 2dyw] \times \\
&\quad [2xz + dv^2 - y^2 - 2duw][z^2 + 2xv - 2yu - dw^2].
\end{aligned}
$$

In the case, $d = 2$, we have $g_2(11, 10, 9, 8, 7, 6) = 1$, which resolves down to the solutions $g_2(3, 2) = 1$ and $g_2(5, 4, 3) = 1$.

10. Pell's equation of general degree

Let n be an arbitrary positive integer and d be an integer with a real nth root θ of the same sign. Define

$$
\begin{aligned}
g_d(x_1, x_2, \ldots, x_n) &= N(x_1 + x_2\theta + x_3\theta^2 + \cdots + x_n\theta^{n-1}) \\
&= \Pi_{i=0}^{n-1}(x_1 + x_2(\zeta^i\theta) + x_3(\zeta^i\theta)^2 + \cdots + x_n(\zeta^i\theta)^{n-1})
\end{aligned}
$$

where ζ is a primitive nth root of unity.

In particular, for a given positive integer k, we have that

$$
\begin{aligned}
N(k^{n-1} + k^{n-2}\theta + k^{n-3}\theta^2 &+ \cdots + \theta^{n-1}) \\
&= \Pi_{i=0}^{n-1}\frac{k^n - d}{k - (\zeta^i\theta)} = \frac{(k^n - d)^n}{k^n - d} \\
&= (k^n - d)^{n-1}.
\end{aligned}
$$

When $d = k^n - 1$, we deduce that $(k^{n-1}, k^{n-2}, \ldots, k, 1)$ is a solution of $g_d(x_1, \ldots, x_n) = 1$. When $d = k^n + 1$, we find that $(k^{n-1}, k^{n-2}, \ldots, k, 1)$ is a solution of the same equation when n is odd and of $g_d(x_1, x_2, \ldots, x_n) = -1$ when n is even.

Let $d = 2$. Then $g_2(1, 1, \ldots, 1, 1) = (-1)^{n-1}$. Since

$$(1 + \theta + \theta^2 + \cdots + \theta^{n-1})^2$$
$$= 1 + 2\theta + \cdots + n\theta^{n-1} + (n-1)\theta^n + \cdots + 2\theta^{2n-3} + \theta^{2n-2}$$
$$= 1 + 2\theta + \cdots + n\theta^{n-1} + 2(n-1)$$
$$\quad + 2(n-2)\theta + \cdots + 4\theta^{n-3} + 2\theta^{n-2}$$
$$= (2n-1) + (2n-2)\theta + \cdots + (n+1)\theta^{n-2} + n\theta^{n-1}.$$

whence $(2n-1, 2n-2, \ldots, n+1, n)$ is a solution of $g_2(x_1, x_2, \ldots, x_n) = 1$.

Let $d = 3$. Then

$$g_3(2, 1, 1, \ldots, 1) = \Pi_{\zeta^n=1}\big(2 + (\zeta\theta) + \cdots + (\zeta\theta)^{n-1}\big)$$
$$= (-1)^n \Pi_{\zeta^n=1} \frac{(1 + \zeta\theta)}{(1 - \zeta\theta)}.$$

If n is even, then $\zeta^n = 1$ if and only if $(-\zeta)^n = 1$, and so $g_3(2, 1, 1, \ldots, 1) = 1$. If n is odd, then when $\zeta^n = 1$, $-\zeta$ is a $2n$th root of unity which is not an nth root of unity. However, for odd $n = 2k - 1$, we can find another solution. It turns out that

$$g_3(3k - 2, 3k - 3, \ldots, k + 1, k) = \Pi_{\zeta^n=1}(1 + \zeta\theta)/(1 - \zeta\theta)^2$$
$$= (1 + 3)/(1 - 3)^2 = 1.$$

Bibliography

1. A. H. Bell, Solution to the celebrated indeterminate equation $x^2 - Ny^2 = \pm 1$, *Amer. Math. Monthly* 1 (1894), 53–54, 92–94, 169, 239–240.

2. Johannes Buchmann, On the computation of units and class numbers by a generalization of Lagrange's algorithm, *J. Number Theory* 26 (1987), 8–30.

3. J. Buchmann, M. Pohst and J. v. Schmetton, On the computation of unit groups and class groups of totally real quartic fields, *Mathematics of Computation* 53 (1989), 387–397.

4. T. W. Cusick, Finding fundamental units in cubic fields, *Math. Proc. Camb. Phil. Soc.* 92 (1982), 385–389.

5. ———, Finding fundamental units in totally real fields, *Math. Proc. Camb. Phil. Soc.* 96 (1984), 191–194.

6. T. W. Cusick and Lowell Schoenfeld, A table of fundamental pairs of units in totally real cubic fields, *Mathematics of Computation* 48 (1987), 147–158.

7. I. Gaál and N. Schulte, Computing all power integral bases of cubic fields, *Mathematics of Computation* 53 (1989), 689–696.
8. K. Mahler, Inequalities for ideal bases in algebraic number fields, *J. Austral. Math. Soc.* 4 (1964), 425–448.
9. M. Pohst and H. Zassenhaus, *Algorithmic algebraic number theory,* Cambridge, 1989.
10. ——, On effective computation of fundamental units I, *Mathematics of Computation* 38 (1982), 275–291.
11. M. Pohst, P. Weiler, and H. Zassenhaus, On effective computation of fundamental units II, *Mathematics of Computation* 38 (1982), 293–329.
12. Hugh C. Williams, Continued fractions and number-theoretic computations, *Rocky Mountain J. Math.* 15 (1985), 621–655.
13. H. C. Williams and G. W. Dueck, An analogue of the nearest integer continued fraction for certain cubic irrationalities, *Mathematics of Computation* 42 (1984), 683–705.

Index